AF528541

Principles and Methods of Orchard Establishment

THE AUTHOR

Prof. Bibhas Chandra Mazumdar had been in the profession of teaching of post-graduate students of several Indian Universities for more than 34 years and supervised researches towards 72 theses of post-graduate students and doctoral fellows. He had also served in fruit development, extension education, farm management and in research positions for more than 7 years. Singly or jointly with others, he has authored 16 books, 163 research papers and 55 non-research articles in horticulture, human nutrition, medicinal plants and plant physiology.

Principles and Methods of Orchard Establishment

2nd Revised and Enlarged Edition

– Author –

Dr. Bibhas Chandra Mazumdar

M.Sc. Agri., Ph.D.

Ex. : Professor in Horticulture,
Institute of Agricultural Sciences, Calcutta University;

Reader in Horticulture,
B. C. Krishi Vishwavidyalaya, Kalyani and Mohanpur;

Lecturer in Horticulture,
Allahabad Agricultural Institute, Allahabad;

Fruit Development Officer, Govt. of West Bengal, Burdwan.

2024

Daya Publishing House®
A Division of
Astral International Pvt. Ltd.
New Delhi – 110002

© 2024 AUTHOR
ISBN: 978-93-5461-989-2

Publisher's Note:

Every possible effort has been made to ensure that the information contained in this book is accurate at the time of going to press, and the publisher and author cannot accept responsibility for any errors or omissions, however caused. No responsibility for loss or damage occasioned to any person acting, or refraining from action, as a result of the material in this publication can be accepted by the editor, the publisher or the author. The Publisher is not associated with any product or vendor mentioned in the book. The contents of this work are intended to further general scientific research, understanding and discussion only. Readers should consult with a specialist where appropriate.

Every effort has been made to trace the owners of copyright material used in this book, if any. The author and the publisher will be grateful for any omission brought to their notice for acknowledgement in the future editions of the book.

All Rights reserved under International Copyright Conventions. No part of this publication may be reproduced, stored in a retrieval system, or transmitted in any form or by any means, electronic, mechanical, photocopying, recording or otherwise without the prior written consent of the publisher and the copyright owner.

Published by : **Daya Publishing House®**
A Division of
Astral International Pvt. Ltd.
– ISO 9001:2015 Certified Company –
4736/23, Ansari Road, Darya Ganj
New Delhi-110 002
Ph. 011-43549197, 8130496929
e-mail: info@astralint.com
Website: www.astralint.com

Printed at : **Replika Press Pvt Ltd**

Dedicated to the venerated memory of the American missionary,

Prof. W.B. Hayes
(1900-1957)

who had an unfathomable love to the soils and plants of India, who had gifted all his possessions for the cause of horticultural education in this country and who spoke his last words on this Earth to his students while teaching in the orchard of the then Allahabad Agricultural Institute.

Preface

A person who does not like fruits is difficult to be met with. This luscious gift of the Mother Earth appeases people of all countries regardless of gender, age and food habits. The health scientists authenticate that fruits of various types are ideal aliments to the human beings in providing nourishment, regaining vitality, warding-off and healing many aliments and restoring normal health.

To provide fruits for consumption of the teeming millions in the world, their production in befitting plantations is necessary. It may be mentioned that India has the potential to provide soil and climatic condition that is required by any species of fruit crop consumable to the human beings. Provision of area having favourable eco-environment is though an essential factor, augmenting fruit production has its reality only when productive capacity of the fruit crops is stepped up in terms of quantity and quality and evidently, this is possible by application of technology.

The study of the scientific production of fruit crops has in fact, two parts, which are firstly, acquiring knowledge on the *principles* of fruit culture and secondly, the *methodology* involved in the culture of individual fruit crops. Under Indian condition, although a number of authentic books are now available that have provided the appropriate technology specifically required for husbandry of the different crops, titles dealing exclusively on the principles and methods involved in fruit culture are not adequate.

On this delibration, the present book has been attempted which has pursued one vital part in the area on principles of fruit culture, *i.e., how to set up an orchard,* in other words, the stepwise procedure to be followed in its execution from the time the idea has cherished in one's mind. The book thus, seeks to meet the requirements of the under-graduate and post-graduate students of fruit science, horticulture and related areas on the one hand and to those, who desire to take up orcharding on the other hand.

As regard acknowledgements, the author gratefully recognizes two categories of people in particular. They are firstly, the fruit growers of West Bengal with whom he had the opportunity to interact when he was a Fruit Development Officer under the West Bengal Government in the nineteen sixtees before taking up researches for his doctoral programme and secondly, the students in his teaching career since January 1971. In fact, the volleys of queries faced from these people and also their valued suggestions have been the basis to take up the present work by the author.

However, the assistance from others in shaping the title in the present form and especially from Mrs. Purabi Mazumdar, M.A., B.Ed., Dr. Kaushiki Mazumdar, M.Sc., Ph.D. and Dr. N. K. Dutta, M.Sc., Ph.D. is also acknowledged with gratitude.

The author will feel amply rewarded if any reader gets any benefit from this monograph. He also appeals the academic persons and the fruit growers to provide valued suggestions to him so that the title could be brought out in a still enriched form in future, atlthough the present edition has been enlorged from that of the earlier edition.

Bibhas Chandra Mazumdar

Contents

Chapter 1

Introduction

1.1 Definition and Concept of Fruit

The term, fruit has been defined in botany as a plant organ, which is formed from the ovary by its growth and in some cases, this organ may remain combined with parts of the flower or inflorescence in mature condition. Botanically designated fruits usually contain one or more *seeds*, which may be either completely or partially developed in them when they are in mature condition, However, the seed may even be totally absent in a fruit. For example, if a fruit develops without fertilization, parthenocarpically as in banana, pineapple seedless varieties of grape etc., or made to do so by applying suitable measures, then also the seedless fruits are grouped under the category of fruits by the botanists. In any case, *it is essential that a botanically defined fruit should contain the ovary*.

In horticulture, on the other hand, the term fruit is used in somewhat restricted sense. In this science, a botanically defined fruit is although considered in many cases as a fruit but not in all. Again, there may be such plant organ which is not recognized as a fruit in botany but is done so in horticulture.

To elucidate this, it is brought to notice that the horticultural scientists lay emphasis on certain conditions or criteria to use the term, fruit. Among those conditions, the following *three* are considered to be of greater significance. Accordingly, a plant organ is designated as a fruit in horticulture when it fulfils the following three requirements. But it also needs pointing out that the conditions are *flexible and are not rigidly followed*.

The three conditions or requirements are elaborated in the following:

1. Botanical Part

In botany, a fruit is considered to be a mature or a ripened ovary. In that sense, fruits of mango, guava, banana, papaya, grape *etc.*, which develop on maturation and ripening of the ovaries are all grouped under fruits in botany. Similarly, in horticulture also, these are considered as fruits.

On the other hand, fruits of pumpkin, brinjal, tomato, lady's finger, gourds *etc.* could be named, which though are mature ovaries and hence, are considered as fruits in botany but these are not considered as fruits in horticulture. In this science, these are regarded as *vegetables*.

In this connection, even a plant organ which has not developed from ovary, viz., *cashew-apple*[1] may be cited as an example which is considered to be a fruit in horticulture. In cashew tree, the peduncle with the thalamus grow by swelling, becomes fleshy, develop sweet-sour taste as well as an attractive yellowish or reddish or a mixed colour and a pleasant aroma. For these attributes, this organ resembles a fruit to a common man. With this consideration, the horticultural scientists have been tempted to call this organ as a fruit and a catchy name has been ascribed to it, i.e., cashew-apple (Figure 1.1). However, the cashew-apple does not contain any ovarian part and hence, it is never a fruit (pseudocarp) according to the botanical norms. Botanically, the actual fruit in cashew is the kidney-shaped nut, known as cahew-nut, which remains attached to the distal end of the cahew-apple.

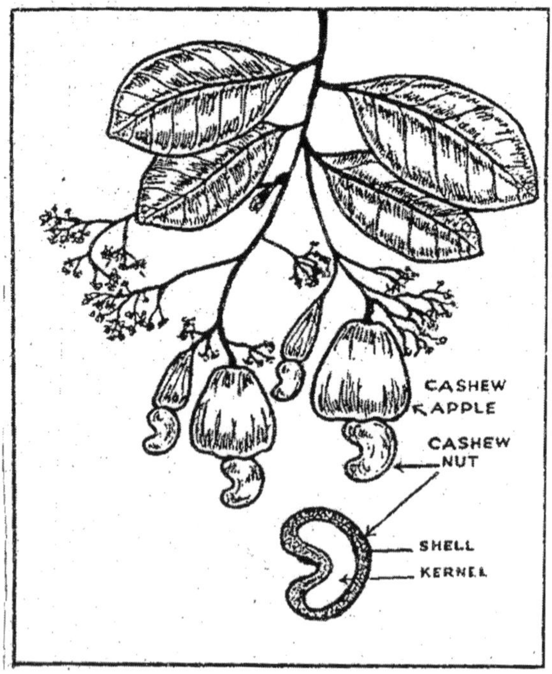

Figure 1.1: Cashew-nut and Cashew-apple.

[1] Cashew-apple is consumed like fruits as raw but more commonly, a fermented drink known as feni is prepared from its juice, especially in Goa and Konkan region and sometimes jam is also prepared from this.

2. Palatableness

Palatableness of a botanical fruit to the human beings is given great emphasis to consider it a fruit in horticulture. That is to say, the horticultural scientists are of the opinion that a botanically defined fruit may be regarded to be a horticultural fruit provided it has an *appealing taste* to the human beings, at least when it is fully mature or ripe. Accordingly, fruits like mango, banana, guava, sweet and mandarin oranges, apple, pear *etc.* well satisfy this condition by their satiable taste and are rightly designated as fruits. But exceptions are also noted at least in case of musk-melon and water-melon. Despite the fact that these fruits are highly pleasant when taken afresh on ripening, these are more commonly grouped under vegetable crops than fruits, particularly by the horticultural scientists and authors of India.

In contrast, fruits of Indian hog-plum, dillenia, karonda, aonla, sour lime, lemon *etc.* are some of many other examples, which lack agreeable taste if taken as raw even in ripe condition but still they are placed in the list of fruits in horticulture.

3. Manner of Consumption

How a botanical fruit is actually consumed by the human beings is considered to be an important determining factor to recognize it as a fruit in horticulture. In other words, the horticultural scientists are of the opinion that a fruit should generally have to be such a food of plant source, that is normally consumed by the human beings in fresh form *i.e.*, without *cooking*. Having considered this criterion, botanical fruits like pumkin, brinjal, gourds, *etc.* have evidently been grouped as vegetables and not fruits in horticultural sense, because of the fact these these are consumed only after cooking and never as raw. On applying the above principles, mango, banana, guava, papaya, oranges, grape, apple, pear *etc.* have also been duly considered as fruits as they are relished by the people in fresh form *i.e.*, without cooking.

But such a pointer of distinction again has its limitation at least in case of water-melon, musk-melon and cucumber. Despite the fact that throughout the world these are taken by people in fresh form, and without any sort of cooking, these botanically designated fruits have been brought under vegetable crops in the horticultural literature.

On contrary, fruits like Indian hog-plum, carambola, karonda, aonla, dillenia *etc.*, may be named, which are hardly possible to be consumed without cooking, or making products from them but disregarding this aspect, these have been included as fruits in horticulture.

Besides the above criteria, maturity of the plant organ is another factor that is paid attention to by many horticultural scientists to call it a fruit or a vegetable. Examples may be cited particularly of jackfruit and papaya. In ripe condition, these are consumed as fresh and are regarded as fruits, while consumption of them in unripe condition is done after cooking and hence, are treated as vegetables.

In conclusion therefore, it can be said that the term, fruit is used in horticultural science in a *loose-sense* and any set of clearly defined norms is hardly maintained by the horticultural scientists to put forth the definition of fruits. The horticultural scientists have in fact, preferred to enlist a number of botanical species of their own choice to include them as members of the fruit kingdom.

From the foregoing pages, it now appears to be a justifiable approach in the area of horticultural science to make a fresh review on the terminology and provide suitable and clearly specified norms to determine which crops should rightly be accepted as fruit crops.

1.2 History of Consumption of Fruits by Human Beings

The historians believe that the fruits being edible substances to the human beings had been a *female invention*. In fact, our pre-historic ancestors were inured to meet up their hunger from the flesh of wild animals by hunting them. With rolling of ages, the animals began to dwindle in forests while human population began to have steep increase, consistently with increase in establishment of family life. The womenfolk who used to stay in the homes unlike the male counterparts first observed that birds and animals used to eat fruits. The womenfolk tasted fruits of some trees and they were charmed with their lusciousness. They further observed that the seeds of the fruits falling on ground gave rise to the same trees and *bore the same kind of fruits.*

They brought this fact to the notice of the male persons of the families and advised them to sow the seeds near their dwelling places. Thereby, they could secure food, *i.e.,* fruits on harvesting from the trees and visiting forests to hunt animals could therefore be dispensed with. The menfolk of the families accepted the novel idea and began to translate it into action. *Fruit culture thus, took its birth* which was many ages before the birth of agriculture.

The Mother Earth remains to be the home of about 240,000 species of plants among which about 5000 species have only been scrupulously examined. More than 500 species of plants do not appear to come under the fruit kingdom including primary or major fruits and less utilized or minor fruits.

A question that is, which species of fruit plants had actually appeared on Earth at the early period and fruits of which species had first been grown and tasted by the human beings has shaken much attention of the scientists who have concentrated on ancientness. Adequate archaeological evidence has unearthed the fact that it was the *fig* tree, which was domesticated first in the Near East about 11,400 years ago, *i.e.,* a thousand years before wheat, barley and legumes were cropped in that part of the world.

Some of the species of fruits which had been started to be grown in the world in ancient time has been mentioned by Bakhru (Ref. Bakhru, H.K. 2007. *Foods that heal-the Natural way to Good Health*. Orient paperbacks, N.Delhi-110002). According to this author, *date* is one of the earliest fruits cultivated by man. It was grown in Mesopotamia more than 5000 years ago. He mentioned that there has been evidence of growing fig in Egypt around 4,000 B.C. and growing of grapes in that country started about 8000 years ago. Mention has been made of growing apricot in China about 4000 years ago and almond in Syria and Palestine since pretty long time. Medicinal properties of bael has been stated to be known to the Indian people since pre-historic times. In Hindu mythology, growing of mango and some other fruits has been amply stated and claimed to be grown in this county since pre-historic time.

1.3 Importance of Fruits to Human Beings

Fruits provide enormous benefits to the human beings. The major importance of this gift of the Mother Earth is discussed below in short.

1. Savouriness

Savour, in other words, palatability is beyond doubt, a very important attribute of fruits. Edward Stevens has quoted, "Some people like non-vegetarian food, some people prefer vegetarian food but everybody is fond of fruits." It may be mentioned that those fruits which are very sour, astringent or have bitterness are not relished in their fresh from but delicious products could be made from many of them.

In this connection, it needs pointing out that such fruits which are very sweet and have negligible sourness do not appeal much. On contrary, those which are much sour and have little sweetness are also disliked. Across the world, it is observed that fruits which have adequate sweetness along with some sourness (acidity) are most liked by majority of the people. In other words, those fruits which have a *desirable blend of sweetness and sourness* and which signifies much sweetness and little sourness make most appeal for their *consumption in fresh from* to majority of the people.

2. Nutrition

Nutrition (Latin, *nutritio* = nourish) refers to the process of taking nourishing food (= nutriments) and their proper use in the body in regard to growth, metabolism and repair of body tissues. The components of nutrients are carbohydrates, proteins, fats, vitamins, minerals and water and the body takes these six chemical components through food and drink. It may be stated that all the above nutritive constituents could be obtained from fruits, in other words, *combination of fruits* belonging to different species. Among all the constituents, fruits are however, most potent food-stuff in providing the requirement of vitamins and minerals in the body. Vitamins and minerals have special significance in protecting the body against diseases and accordingly, fruits and vitamins are called *protective food*.

In the following has been mentioned some fruits which are relatively rich sources of nutritive components, obtaining the values from various books on fruit science and particularly from the report of the National Institute of Nutrition, Govt. of India (Ref. Gopalan C., Rama Sastri, B. V., Balasubramanian, S.C. and revised by

Narasinga Rao, Deosthale, Y.G. and Pant, K.C. 2018. *Nutritive Value of Indian Foods.* N.I.N., Hyderabad – 500 007).

(a) Carbohydrates

As compared to cereals, millets, pulses and oilseeds, fruits are not rich sources of carbohydrates, proteins and fats. Nevertheless, there are some fruits and particularly nuts which contain fairly high amount of these components.

Carbohydrates are principal source to provide *energy* in the body, besides *dietary fibres.* They also take part in various metabolic reactions. Some fresh and dried fruits which contain fairly high amount of carbohydrates are stated in Table 1.1

Table 1.1: Carbohydrate Content (gram per 100 gram) of some Fruits

Dried date	75.8 g
Black currant	75.2 g
Raisin	74.6 g
Durian	34.1 g
Bael	31.8 g
Banana	27.2 g
Diospyros	26.6 g
Custard apple	23.5 g

(b) Proteins

Proteins are indispensable for growth, especially of children, synthesis of enzymes, hormones, antibodies and haemoglobin, insulin, collagen, fibrin, fibrinogen, immunoglobin *etc.* Fruits and nuts having relatively high amount of protein have been stated in Table 1.2.

Table 1.2: Protein Content (gram per 100 gram) of some Fruits and Nuts

Cashew kernnel	21.2 g
Almond	20.8 g
Pista nut	19.8 g
Walnut	15.6 g
Water-chestnut	4.7 g
Coconut	4.5 g
Manila tamarind	2.7 g
Jackfruit	1.7 g

(c) Fats (Lipid matter)

Fats make vitamins A, D, E and K soluble such that, they are rendered acceptable to the body. Fat is also a *rich source of energy*. It is necessary for growth, nerves, reproduction, lactation and insulation of the body. Table 1.3 has provided list of some fruits and nuts which contain high amount of fat.

Table 1.3: Fat Content (gram per 100 gram) of some Fruits and Nuts

Walnut	64.5 g
Almond	58.9 g
Pista nut	53.5 g
Cashew nut	46.9 g
Coconut	41.6 g
Avocado	22.8 g
Durian	3.9 g
Elephant apple	3.7 g

(d) Vitamins

The term, vitamin is derived from the Latin word, *vita* which means life and this word is suffixed with the word, amine, which denotes nitrogenous compound (amino acid). Such a name was proposed in consideration to the fact that the vitamins were amino acids in chemical form. But the term is unfortunate as it was learned later that all vitamins are not amino acids. Vitamins are required in the body in small and in some cases, in very small amount but that amount becomes indispensably needed for nourishment in the body. Some of the vitamins have therapeutic properties as well when used in appropriate amount, which is apart from the nutritive role of them.

There are in fact, various types of vitamins and each one has its specific activity in the body. According to solvents in which they are soluble, the different vitamins are conventionally classified into two broad groups. These are, (i) fat (or, fat-solvent) soluble vitamins *e.g.*, vitamins A, D, E and K, and (ii) water soluble vitamins, which consist of the vitamins under the B-group (known as B-complex vitamins) and the vitamin C. (Vitamens are compounds which show similar biological activity as with the respective vitamin).

Fruit analytical studies done in different parts of the world made it clear that majority of the horticulturally recognized fruits are rich or fair sources of one or more of the vitamins. It also needs mentioning in this context that a large number of fruits are taken as raw, *i.e., without heating and cooking.* Accordingly, some of the vitamins which are easily lost on heating, *e.g.* vitamin C and vitamin B_1 are saved from destruction and thereby, the entire amount of these vitamins which is present in them are utilized in the body.

(i) Vitamin A

Vitamin A is not present in any fruit. Fruits contain the pigment, known as *carotene* which is converted into vitamin A (Retinol) on entering into the body. There are different isomeric forms of carotene. Among these, *beta*-carotene is most important and one molecule of *beta*-carotene gives rise to two molecules of vitamin A. Deficiency of vitamin A in the body is responsible for night blindness, drying and abnormality of cornea and retina and results in blindness gradually. Fruits and nuts which are rich in carotene have been stated in Table 1.4.

Table 1.4: Carotene Content (International Units per 100 gram) of some Fruits

Ripe mango pulp	4800 IU
Ripe papaya pulp	2020 IU
Persimmon	1710 IU
Dried date	600 IU
Jackfruit	540 IU
Mandarin orange	350 IU
Fig	270 IU
Carambola	240 IU

(ii) Vitamin E

This vitamin which goes by the name as tocopherol (syn. Tocotrienol) protects red blood corpuscles against destruction and is helpful in eye disorder (retinal myopathy), skin health, nervous disorder and is an anti-oxidant. Table 1.5 presents the vitamin E content of some nuts and fruits.

Table 1.5: Vitamin E Content (milligram per 100 gram) of some Fruits and Nuts

Almond	26 mg
Pecan nut	24 mg
Hazel nut	15 mg
Brazil nut	7.9 mg
Pista nut	2.3 mg
Black berry	1.2 mg
Ripe mango	1.1 mg

(iii) Vitamin B_1

This vitamin which goes by the name as thiamine (syn. *Aneurine*) prevents and cures the beriberi disease and is of much help in nervous system. Fruit and nuts which are rich sources of this vitamin have been stated in Table 1.6.

Table 1.6: Thiamine Content (International Units per 100 gram) of some Fruits and Nuts

Cashew kernel	630 IU
Walnut	450 IU
Almond	240 IU
Banana	150 IU
Apple	120 IU
Grapefruit	120 IU
Mandarin orange	120 IU
Red plum	120 IU

(iv) Vitamin B_2

Deficiency of this vitamin (syn. Riboflavin, Lactoflavin, Ovoflavin) causes anaemia, cataract, skin disease, lethargy, cheilotic fissure, glossitis, keratitis *etc.* Table 1.7 presents the list of some fruits which are relatively rich sources of riboflavin.

Table 1.7: Riboflavin Content (International Units per 100 gram) of some Fruits and Nuts

Bael	1191 UI
Ripe papaya	250 UI
Cashew kernel	190 UI
Elephant apple	170 IU
Litchi	123 IU
Pineapple	120 IU
Pomogranate	100 IU
Mandarin orange	60 IU

(v) Niacin

Deficiency of this B vitamin (syn. Nicotinic acid) causes pellagra disease, diarrhoea, anaemia, drying of skin *etc.* Table 1.8 has listed some fruits which are rich sources of niacin.

Table 1.8: Niacin Content (International Uniter per 100 gram) of some Fruits and Nuts

Amond	2.5 IU
Cashew kernel	2.1 IU
Walnut	1.6 IU
Bael	0.9 IU

Dried date	0.6 IU
Fig	0.5 IU
Jackfruit	0.4 IU
Grapes	0.3 IU

(vi) Vitamin C

Vitamin C (syn. Ascorbic acid) has plentiful physiological actions in the body. It is necessary for synthesis of *collagen protein* and acts as a co-factor in many enzyme systems. Among a number of therapeutic effects, it prevents and causes scurvy disease, gout, dental caries, iron deficiency anaemia, allergy, throat infection *etc.* It has *strong anti-oxidant properties.* Fruits which are rich sources of vitamin C have been stated in Table 1.9.

Table 1.9: Vitamin C Content (milligram per 100 gram) of some Fruits

West Indian cherry	1700 mg
Aonla	700 mg
Ripe guava	212 mg
Cashew apple	180 mg
Monkey jackfruit	135 mg
Manila tamarind	108 mg
Ber	76 mg
Sour lime	63 mg

[Cacado plum (not grown in India) contains much higher amount]

(e) Minerals

These are naturally occurring chemical elements which exist as inorganic elements or salts or components of organic compounds. According to *relative amount* of them that are taken up by the body, they are classified as macro- or major- minerals that are required in relatively high amount and micro-minerals which are also known as *trace-elements* and are required in small or even very small amount. Minerals are required for various physiological functions in the body, such as co-factors of enzymes, maintaining electrolytic balance, osmotic pressure, transportation *etc.*

Quite a large number of macro- and micro minerals are taken up by the human body. Among these, calcium, phosphorus and iron have particularly great significance. Calcium is indispensable for formation of bones and teeth, muscular movement, cardiac functioning, blood clotting *etc.* Phosphorus is necessary for utilization of calcium and in combination with calcium, it protects bones, teeth and muscles. Iron is indispensable for haemoglobin formation.

Table 1.10 has provided the list of some fruits which are relatively high sources of calcium, phosphorus and iron

Table 1.10: Calcium, Phosphorus and Iron Content (milligram per 100 gram) of some Fruits and Nuts

Calcium		*Phosphorus*		*Iron*	
Almond	230 mg	Almond	490 mg	Currant	8.5 mg
Pista	140 mg	Cashew kernel	450 mg	Water melon	7.9 mg
Walnut	100 mg	Pista	430 mg	Pista	7.7 mg
Cashew kernel	50 mg	Walnut	380 mg	Raisin	7.7 mg
Currant	130 mg	Coconut	240 mg	Dried date	7.3 mg
Elephant apple	130 mg	Currant	110 mg	Cashew kernel	5.8 mg
Phalsa	129 mg	Elephant apple	110 mg	Almond	5.1 mg
Dried date	20 mg	Raspberry	110 mg	Custard apple	4.3 mg

It may be stated that many other elements play great role in the body. Some examples may be cited as sodium, potassium, magnesium, sulphur, chlorine, iodine, manganese, copper, zinc, chromium *etc.* Small or adequate amount of these mineral elements are found to be existent in fruits and nuts belonging to many species.

In regard to nutritive value of fruits, a report may be cited (Ref. *The Telegraph* Daily, Calcutta, Monday, May 7, 2018). It states, "Researchers who looked at more than 5,500 women from Britain,, Ireland, Australia and New Zealand found that those who consumed the least fruit were 50 per cent more likely to be infertile. And compared to women who ate fruit three or four times a day in the month before conception, women who ate fruit less than one to three times a month took half a month longer to become pregnant."

3. Medicinal Value

Apart from the nutritive value, a great many horticultural fruits are known to possess a number of medicinal properties as well and some of these have been stated in the following.

Mango

Ripe mangoes being richest source of beta-carotene among all fruits are of great help in some types of eye disorder. They are also helpful in hepatitis, arthritis, haemoptosis and nervous disorders.

Banana

In containing ACE inhibitor, it is highly beneficial in hypertension. Besides, in piles, dysentery, constipation, sore-throat and tuberculosis also, the fruits are helpful.

Citruses

All citrus fruits are antidotes in scurvy, dyspepsia and malarial fever. Juice of mandarin orange is astringent, anti-vomiting and aphrodisciac, juice of sweet orange is anti-hypertensive and useful in bilious affliction, juice of shaddock is useful in jaundice and influenza, juice of lemon has anti-pneumonic and anti-rheumatic factors, while juice of sour lime counters bilious vomiting.

Guava

Being rich source of vitamin C and pectin, semi-ripe fruits of guava are of great use in scurvy disease and bowels movement respectively.

Papaya

Ripe fruits are rich sources of *beta*-carotene. Unripe fruits have *papain* thereby, are helpful in digestion of protein, curing of ulcer and is used even in cancer.

Grape

Berries are useful in anaemia, gallstone, stomatitis, cough, wasting diseases and have diuretic properties.

Apple

Fruits of apple are laxative, increase memory power and are helpful in hypertension-induced paralysis of organs. Besides, in anaemia and headache also, they render benefit.

Pineapple

In containing *bromelain*, they digest proteinaceous food easily like unripe papaya and have great role in removal of intestinal worms.

Litchi

Litchis relieve Relieves body ache, increases memory power and is a good appetizer.

Bael

Fruit is highly efficacious in chronic amoebiasis, besides piles, diabetes, stomatitis, diarrhoea, gastrointestinal ulcer and movement of bowels. Being richest source of riboflavin among fruits, ripe fruits are of much help to maintain healthy eyes, skin and mucous membrane.

Jamun

In dyspepsia, stomatitis, loss of appetite, gastrointestinal ulcer, diabetes and diarrhoea, ripe fruits are very much useful.

Aonla

Being the second richest source of vitamin C among all fruits, the fruits of aonla are extremely helpful in scurvy. They are also highly efficacious in jaundice, dysentery, dyspepsia, constipation, cough and infections in vulva and in many other health disorders.

Pomegranate

Juice is anti-pyretic and is a valuable substance for some types of heart diseases.

Date

Dates are known to cure some types of anaemia and also asthma, gonorrhoea, squeezed bladder, besides having aphrodisciac properties.

Almond

Kernels are helpful in ear troubles and deafness, stammering, bone and tooth disorder.

Elephant apple

Fruits have anti-fermentive, anti-pox and anti-malignant properties and are also valued for pestilent fevers.

Pear

In piles, fever, colitis and scurvy, the pear fruits have beneficial effect.

Cashew-nut and Walnut

Kernels have great role in some types of nervous disorders

1.4 Less Costly Food-stuff

Those fruits which are consumable as raw, *i.e.*, without processing or cooking, no further expenditure is actually involved to render them consumable after purchasing from the market. Accordingly, the cost of these fruits as food-item may be much on the lower side than other food-stuff that need to be cooked for consumption.

1.5 Use in Industries

Fruits of many species are utilized in a number of industries also. The fruit preservation and processing industries could be mentioned first, where a wide range of fruits serve as raw materials.

In other industries as well, requirement of some particular species of fruits becomes necessary. For example, the dried milky latex, tapped from immature papaya fruits (which on purification yields the product, *papain*) contains a large number of proteolytic enzymes and hence, becomes necessary in pharmaceutical, textile, wool, distillery, tanning, cosmetic and many other industries. Similar type of proteolytic enzymes obtained from pineapple and fig, which are respectively known as *bromelain* and *ficin* are also used in the pharmaceutical industries. In the same industries, other fruits like citruses, bael, guava, strawberry, raspberry *etc.*, or peels of citruses are required for extraction of pectin, vitamins, flavouring compounds and other organic substances.

Some fruits like olive, apricot, walnut *etc.* have their utilization in extraction of oil for edible or cosmetic use. Cashew-nut shell liquid (CNSL) extracted from the shells of cashew-nut is another valuable material for many industrial uses.

In some industries, leaves of some fruit trees have also been on use. Examples may be cited as of papaya, the leaves of which are required for extraction of *carpaine* (an alkaloid having usefulness as a heart stimulant and amoebicide, mulberry, the leaves of which are required for rearing silkworms, diospyros for making *bidis* (country made cigars), date and palmyra-palm for making fans or other decorative objects and the like.

In some countries, the milky latex tapped from the bark of sapota trees yields *gutta percha*, a commercial product, which is used in the manufacture of chewing gum, dolls and toys. The gummy material obtained from the jackfruit tree is utilized in some rural cottage industries in India. Mention may also be made to the alcohol, ancorin, extracted from custard apple, the potentiality of which appears to exist for use in biopesticide industries.

1.6 Other Importance

Among other importance of fruits, it may be stated that they comprise an important food-stuff to the section of people who live exclusively on vegetarian food. Thus, they have a great role in a country like India, where a large percentage of the population is solely vegetarians.

In this country, fruits are even regarded as sacred food especially by the Hindus and hence, they are valued in offering oblations to the gods and goddesses. Even unused parts of many fruits, *e.g.*, husks of walnut, palmyra, seeds of apple, custard apple, sapota, peach, plum *etc.* leaves of date, palmyra *etc.* are also utilized by craftsmen in many parts of India in making garlands, decorative objects and household materials.

In addition to the above, fruits have their importance in wild-life preservation for, they comprise as the only or the major part of food to the birds and other animals belonged to many species.

1.7 Importance of Fruit Growing in India

The discussion made in the foregoing pages has made it apparent that fruits serve as important nutritive and protective food for human beings and especially for the people of India where a large section of the population lives on vegetarian food. This obviously necessitates *large-scale production of fruits* in this country.

However, production of fruits in this sub-continent in abundance has other socio-economic reasons as well and the notable ones could be mentioned as follows.

1. In Alleviating Food Problem

A perusal on the food consumption pattern in India points out that in general, the people in this country are in the habit of consuming higher quantity of cereal food than what is actually recommended. A survey made has provided evidence to the fact that in some parts of the country, a section of people doing moderate physical work consume as high as 700 gram or more of cereal per head per day as against the recommended schedule of 400 gram.

Such wanton consumption of cereals by the people in this country could hardly get support for two reasons. These are, firstly, by fulfilling the stomach with super-optimum quantity of cereal food consumption of other type of food obviously becomes lower. Accordingly, the people are chronically subjected to deficiency of the nourishment that would have been derived from other food-stuff. Secondly, an abundant intake of cereals by the people is causing to exert an *undue pressure on the cereal stock of the country*. It is also noticeable that the people belonging to different regions in the country are usually accustomed to consume particular type of cereal only and the demand is felt greater for cereals like rice and wheat and thereby, much *greater pressure* is in fact, fallen on these cereals.

The high rate of consumption of cereals particularly by the ordinary working class of people in the country is however, possible to be lessened much, if part of such food is made to substitute by some suitable *supplementary food* in their daily diet.

In choosing the supplementary food for partial replacement of the extra cereal food in the diet, the factors that are associated with high rate of cereal consumption by the Indian people should not be disregarded. In other words, the supplementary food needs to be chosen with consideration to the fact that it is merited with three special attributes in particular. These are firstly, it should be capable to provide adequate *bulk* in the diet as may satisfy the hunger of the people, secondly, it should be of good *calorific value* as may provide adequate energy on its consumption and thirdly, the *nutritional value* should also be high.

Taking into consideration the above factors, it may be opined that the various types of horticulturally listed *fruits*, which are grown in this country could serve as the ideal supplementary food in reducing the over-consumption of cereal food by the Indian people.

Fruits as potential sources of bulk food supply is evident from the fact that the yield as food per unit cropped area of land in case of many of them greatly surpasses the major and the minor cereals (Table 1.11). It also needs mentioning that even with supply of lesser agro-inputs, the yield of many fruit crops is found to be much higher than the cereal crops.

As compared to the cereal food, the common fruits are although of low calorific value, some fruits can be mentioned which on consumption provide fairly high amount of calorie. Besides, if the energy value of fruits harvested per unit cropped area is compared with the energy value of cereals, many of the fruits are found to surpass the cereals (Table 1.12).

Table 1.11: Comparative Yield of some Fruits and other Agronomic Crops in India

Crop	*Average Yield (tonnes/hectare)*	*Crop*	*Average Yield (tonnes/hectare)*
Paddy	1.30-1.55	Guava	6.5-11.0
Wheat	1.80-2.00	Sweet orange	10.0-18.0
Maize	1.15-1.46	Mandarin orange	10.0-18.0
Sorghum (Jowar)	0.60-0.75	Grape	16.0-20.0
Bulrush millet (Bajra)	0.30-0.65	Pineapple	12.5-28.0
Pulses	0.50-0.60	Jackfruit	198.0-252.0
Potato	13.0-16.0	Apple	4.5-7.0
Mango	8.0-12.5	Pear	10.0-15.5
Banana	18.5-28.5	Peach	5.7-7.1
Papaya	15.0-25.5	Plum	6.8-8.1

Table 1.12: Comparative Energy Value of Wheat and some Fruits Harvested from one Hectare Area of Land

Crop	*Energy Value per Hectare Area of Cropping*
Wheat	5,623.090 K Cal (23,616,978 KJ)
Date	22,390,960 K Cal (94,042,032 KJ)
Banana	19,336,140 K Cal (81,211,788 KJ)
Papaya	15,000,000 K Cal (63,000,000 KJ)
Guava	22,390,960 K Cal (94,042,0322 KJ)

High nutritional value of fruits needs no emphasis and has been indicated in the earlier part (Tables 1.1 to 1.10)

The above discussion therefore, clearly implies that fruits may be regarded to be the *appropriate supplementary food* for reducing to a considerable extent the high rate of consumption of cereals by the Indian people and thereby, lessen the *over-pressure*, which is being shouldered by the cereal stock of the country at the present time. Another notable feature in favour of fruits lies in the fact that they are acceptable to the people, belonging to both the vegetarian and non-vegetarian food habits.

The importance of growing fruit crops to ensure higher production of fruits may therefore, prove to be an effective way in alleviating the food problem in the country by discouraging the habit of faulty intake of some food-stuff by a section of the population.

2. Generating Rural Employment

The National Commission on Agriculture has pointed out that the agricultural sector possesses maximum potential for employment of the rural people in the country. Again, highest man power utilization in this sector is achieved in farming with fruit crops than with other agricultural crops. It may be cited that while for the principal or main cereal crop like wheat, the average man-day requirement in India is usually not more than 143 per hectare, for fruit crops, it is generally estimated to be 855 per hectare and in case of grape, the figure may reach even upto 2510 for the same area of a vinery.

There has also been a significant growth of cottage- and small-scale industries in the rural sector in many parts of India by the utilization of fruits, fruit parts or different parts of fruit plants. In the cashew-nut processing industry alone, more than 3 lakhs of people are employed in the processing factories in the country, numbering about 600. Fibre production from pineapple leaves is also gaining importance in the southern and eastern parts in the country, thus making avenues to emerge new industry in the rural sector.

The sugary juice obtained by tapping palmyra-palm and wild date is processed in the village industries to make jaggery, sugar candy, fermented drink *etc.* The womenfolk in many rural parts of the country earn their livelihood by dehydrating, processing and making products from many fruits *e.g.*, aonla, ber, mango, banana, jackfruit, bael, elephant apple *etc.* Even seeds or stones of some fruits like apple, custard apple, sapota, peach, plum, palmyra *etc.* serve as materials to make garlands and decorative objects for the inland and foreign markets.

3. Feeding Large- and Medium-scale Industries

Fruit growing has its requisiteness in the running of a host of large and medium-scale industries in India, where fruits serve as the only or one of the raw materials.

The fruit preservation and processing industries could be named first, in the running of which, supply of fruits is inevitable. As a matter of fact, fruits belonging to many species are utilized, or are possible to be utilized in this country. It is heartening to note that the growth of the fruit processing industry in India was although slow upto the early seventies, it has picked up later and has made a promising stride after 1990.

Shelling and processing of cashew-nuts could be looked upon as another horticultural industry in the country, which has its subsistence on the production or availability of cashew-nuts as raw materials. There are about 600 cashew-nut processing units in India with a work force of more than 3 lakhs comprising mostly womenfolk and the estimated installed capacity is about 5 lakh tonnes of raw nuts per annum.

The papaya culture has its significance not only for harvesting edible fruits, but also for production of *papain*, which is the purified from of the milky latex, tapped from the immature papaya fruits. The latex is an abundant source of proteolytic enzymes and hence, the dried form of it, *i.e.*, crude papain or the purified form, the papain finds use in a large number of big or small industries, important of which are pharmaceutics, distillery, textile, tannery, wool, cosmetics *etc.* as stated under the Heading, 1.5.

The industries concerned with production of cooking or industrial oils in the country make utilization of a number of fruits, *e.g.*, olive, apricot, almond, walnut, hazel-nut, cashew-nut, custard apple seeds *etc.* for extraction of oil and accordingly, necessitates culture of these fruits.

Mention may also be made to the fact that a number of fruits like bael, jamun, oranges, lemons, pineapple, strawberry, raspberry, leaves of papaya *etc.* are required by the pharmaceutical or perfumery industries for medicinal preparations, extraction of alkaloids, aromatic compounds, bitter principles *etc.*

4. Earning Foreign Exchange

Production of fruits, nuts and fruit products has been an important step to earn large amount of foreign money. Apart from cashew kernels which are leading foreign money earners in India, mangoes, grapes, dried fruits, pickles *etc.* have also gained ground as export materials to many countries.

5. Profitability

Intelligent selection of the fruit crops to be grown in a locality and adopting scientific cropping of them ensures high economic return to the growers. Table 1.13 provides data on the economics of production of some fruit crops under Indian conditions on a general basis.

Table 1.13: Economics of Production of Mango, Papaya and Grape per Hectare Cropped Area in India

Fruit Crop	*Variety*	*Yield*	*Production Cost (Rs.)*	*Gross Income (Rs.)*	*Net Profit (Rs.)*
Mango	Amrapalee	38,400	20000.00	115,200	95,200
	Mallika	17,200	20000.00	51,600	31,600
	Dusheri	12,400	20000.00	37,200	17,200
Papaya	Pusa Delicious	102.50	20000.00	102,500	82,500
	Pusa Dwarf	78.50	20000.00	78,500	58,500
	Pusa Majesty	95.00	20000.00	95,000	75,000
	Pusa Nanha	63.36	20000.00	63,360	43,500
	Ranchi	34.50	20000.00	34,500	44,000

Grape	Beauty Seedless	15.00	20000.00	75,000	55,000
	Perlette	20.00	20000.00	10,000	80,000

6. Land Utilization

Productive utilization of such type of land, which usually has lesser suitability to grow agricultural crops is another merit in favour of fruit growing. There are a number of fruit crops, which thrive and produce in dry land, adverse soil conditions or irregular terrain, where other crops generally fail to grow or yield satisfactorily.

It needs mentioning in this connection that about 70 per cent of fruit crops are grown in India under dry land conditions, where precipitation is received in two or three months in the year only and have long dry spell. Under such soil moisture deficit conditions, some fruit crops, *e.g.*, custard apple, pomegranate, date, aonla, fig, jamun, elephant apple, ber, carambola, bael, cashew, wild date palmyra, phalsa, monkey jackfruit, cordia *etc.*, are grown well in the country and reasonable production is obtained in absence of other handicaps.

In many parts of the country, lands which are uncultivable, or of inferior soil type are made utilized in raising fruit crops. Some examples may be cited as viticulture in the Madurai district of Tamil Nadu, custard apple in the *chalka* soil in Andhra Pradesh and culture of mandarin orange in the mountainous region in the eastern States. Cashew growing in the sand hills of the western coast, pineapple cultivation in the hill slopes of the eastern States, cashew plantations in the coastal saline belts in the country may be cited as other examples. Even ponds or ditches where pisciculture is not done, a nut crop, *viz.*, water-chestnut is grown in eastern and other parts of India, which ensures high economic return to the growers.

Utilization of spare lands in the urban areas, demesnes, places of public interest, school compounds, along highways, railway tracks, strip plots in rural areas *etc.* by planting fruit trees is also a common practice in this country.

1.8 Objectives to Set Up Fruit Plantations

Fruit plantations are set up with some specific objectives and the common ones under Indian condition may be stated as follows:

1. Commercial

When the fruit crops are grown in an orchard exclusively to earn *profit* by disposing the harvested produce, it is referred to as a *commercial orchard*. Disposal of the produce, *i.e.*, the ripe or unripe fruits harvested from this orchard is done either for use as fresh fruits, or for supplying to the processing factories to make preserved products out of them. An orchard owner may market the produce by himself, in groups or through a co-operative society. In many parts of India, there is also a trend to lease the orchard to a contractor, which is locally known as *patta* system. The lease is usually given to the contractor when the fruits have attained at least one-fourth maturity which the contractor observes and in accordance with, the price is settled. Selection of a contractor through auction basis is also done by many orchrdists in India.

It also needs mentioning that in order to secure an extra income, or to avoid the risk to earn optimum profit, that is feared to happen in some years, *filler trees* with temporary fruit plants or *inter-crops* are sometimes grown in a commercial orchard. The disposal of the harvested produce of these secondary crops is usually done in the same manner as with the principal fruit crops. Orcharding in commercial-scale is however, of more common occurrence under Indian conditions.

2. Private Use

An orchard may be set up by an individual, members of a family, a community, an educational institution, an agency, a company, or any such body with the idea to harvest the fruits for the purpose of consumption among the members, *i.e.*, for private use.

A fruit preservation and processing concern also sometimes maintains its own orchard to get the raw materials. The normal practice is though to purchase the raw material fruits from other persons, orchrdists or agencies, some companies prefer to maintain their own private orchards, particularly to have an assured supply of the raw materials, either partly or fully from their own source and without depending on others.

3. Experimental

The objective of such an orchard is to conduct experiments on fruit crops. These orchards are established and maintained by the research institutes or stations, agricultural or horticultural colleges, universities *etc.*, with the idea that the experiments could be done under controlled condition. As conducting experiments on fruit crops is the primary objective, profit is not of much concern in such an orchard, although expenditure may go very high depending on the experiments that are taken up.

4. Germplasm Collection

For the purpose of *introduction and maintenance* of fruit crops belonging to different species, varieties, cultivars, types, strains *etc.* these orchards are set up by the research institutes, agricultural or horticultural colleges, governments *etc.* These orchards usually consist of adequate space as plant introduction remains to be a continuous process. Even many wild species or types may also become necessary to be introduced. Great care is no doubt, taken for maintenance of the inmates in ideal condition and this may necessitate high cost.

5. Teaching, Demonstration and Training

These orchards are set up for practical teaching of orchard operations and activities to the under-graduate or post-graduate students of fruit science, or imparting training to the horticultural personnels for refresher course, development officers, trainees, orchard owners, nurserymen *etc.* and are owned by agricultural or horticultural colleges, universities, government departments, training centers *etc.* For the purpose of demonstration and identification, *museum orchards* are also maintained which consist of growing one or more fruit trees or shrubs of different species, varieties, strains, types, hybrids *etc.* in rows in a block of land.

6. Propagation

Many nurseries also maintain their own orchards for procuring ready-at-hand the propagating materials like seeds, cuttings, layers, scions or stocks. Such an orchard becomes smaller in size as the species or varieties of fruits desired to be propagated are only maintained in it. Great care is no doubt, taken on health of the fruit trees and their vegetative growth.

7. Controlling Soil Erosion

With the objective to control soil erosion along rivers, canals, dams, seas *etc.*, hilly terrain, or in undulated land, sometimes plantations of some specific types of fruit crops which have good soil-binding capacity are raised. (It is mentionable in this connection that the Portuguese settlers in India first brought cashew in this country in the 16th century for planting in the Malabar coast of the Arabian sea to control soil erosion by the sea, however, without knowing the fact that the nuts of cashew bear so delicious and nutritious kernels in them.)

8. Utilization of Inferior land

For productive utilization of inferior land, plantations are also raised of some hardy fruit trees or shrubs as stated earlier in this Chapter, under the Heading, 1.7 (6)

9. Afforestation

For planting in forests, some fruit trees notably, jamun, jackfruit, monkey jackfruit, diospyros, buchanania, madhuca, sterculia,palmyra palm, walnut, filberts *etc.* are sometimes preferred. Besides, in social or agro-forests also emphasis is given to include some fruit trees.

1.9 Types of Fruit Farms

In consideration to the size, location, types of fruit crops grown, system of culture adopted and other factors, plantations of fruit crops are usually designated by different terms. These have been stated in the following.

The term, *orchard* is by far, the most common one, which is conventionally used to refer to an area where fruit crops like mango, oranges, guava, papaya, apple, pear *etc.* are cultivated. In an orchard, intensive cropping method may or may not be applied.

Plantation is usually referred to a fairly large area, where cropping is done with some particular types of fruit crops notably, banana, cashew, pineapple, date, papaya *etc.*

The term, *field* is somewhat interchangeably used with plantation and is spoken of particularly when *qick-growing fruit* crops, *i.e.*, banana, pineapple, strawberry, melons, cape-gooseberry *etc.* are usually cropped in it.

Estate is a term that had been in vogue in olden days and by such people who used to maintain relatively large area of land for growing fruit crops along with other agricultural or horticultural crops. The term is rarely used nowadays.

The term, *garden* is also sometimes used to refer to a fruit farm, which is perhaps to lay emphasis to the fruit crops grown there with sophisticated agro-practices.

The horticulturists have also a preference to refer to areas of some specific fruit crops with names linking with the names of the fruit crops that are grown there. Thus, an area which is cropped with grapevine goes by the term, *vineyard*, or infrequently as a *vinery*, an area under pineapple crop is referred to as a *pinery* and an orange orchard is given a favourite title, *i.e., orangery. Amra-kunja* is a term which is used by many mango growers in West Bengal for their favourite mango orchards.

Despite the fact that several terms have been on use, *orchard* is no doubt, the common term applied throughout the world. In the present text, the term, orchard that is of common usage has been largely use to refer to areas where culture is done with fruit crops that are of longer duration.

At the present time, some orchardists in India feel interested to use their orchards not only for growing fruit crops but for some other purposes also although the trend is not largely followed. This is usually done to avoid the risk of growing fruits and thereby, to *earn supplementary income from the orchards.* The practice is however, adopted where the size of the orchard is *considerably large*.

The activities that are considered to be taken up in the orchard may be more than one and making suitable arrangement for picnicking is more common. Establishment of children's school is done by some orchardists, naming it as *garden school, orchard kindergarten, crèche etc.* Few orcharidists also consider to build up small house at the fringe area of the orchard for use as resort and naming as *eco-friendly resort.* This aspect has been discussed under the Chapter 20.

Chapter 2

Planning of an Orchard

2.1 Principles and Objectives

An orchard is set up for many years of its existence. High initial cost is also an essential part in its establishment. Further, high expenditure and efficient management become necessary in its running condition. These are the factors which cannot be overlooked when it is planned to set up the orchard and evidently, these should be attended to with a realistic approach rather than from a theoretical viewpoint only. With this perspective, the more relevant factors that are needed to be perused in setting up of an orchard has been dealt with in the following.

1. Necessity to set up orchard

Is there any real necessity to set up the orchard ? Beyond doubt, this is the first and the most pertinent question that demands satisfactory answer before any action is taken up. It goes without saying that in some cases, *e.g.*, in an agricultural or a horticultural college, a training or a research centre, setting up of orchard by the concerned authority is an indispensable requirement for the purpose of teaching, demonstration, conducting experiments *etc.* as the case may be. In other cases, such as for a fruit preservation factory to secure raw materials, or for a nursery, or for other organizations, own orchard may though be of advantage, whether it is absolutely necessary or not needs to be ascertained.

A more critical process of reasoning is however, felt in the establishment of an orchard for commercial purpose. As stated earlier, establishment and maintenance of an orchard necessitate high cost and efficient management. Above all, risk factors may also be there in this avocation due to natural calamity, production principles, demand in the market, transport of the harvested produce, social factors and other conditions. As a matter of fact, it can be worthily opined that orcharding is a highly profiteering enterprise if it runs in the proper way and in absence of that, it may be a failure enterprise.

The above facts should not be set at defiance when a commercial orchard is planned to be built up. It also needs pointing out that the would-be orchardist should have a *real interest* in orcharding and the avocation should not be taken up just out of whim. He should also have the temperament to visit his orchard regularly and feel pleasure in that[2] It is even highly desirable if he himself takes part in some activities of the orchard.

Sometimes it comes to notice that some people are allured to set up orchards to amass money without much investment. They are guided by the supposition that expenditure involved in the orchard is not much in its running condition. Some people intend to set up orchards only on hearing from others that some particular persons earn high profit from their orchards. People of this category are easily motivated by other persons and the consequence may be unfortunate if the latter persons give such advice without having sound knowledge of fruit growing, or have only a smattering. On having these advices directly or indirectly, many imprudent persons take up orcharding. Large fund is invested by them to set up orchards and often they take much interest in the orchard activities during the first few years.

However, in course of time, they often lose interest in orcharding and feel shaky even with a trifling constraint and are reluctant to face it out. For undesirable events, many of such orchardists want to get rid of the avocation and consider of selling their orchards. They form the notion that profit is difficult to obtain from an orchard unlike the loss and propagate this belief to others.

It may be mentioned that spreading such despairing concepts to other persons may have an undesirable impact in that, those who have ardent zeal to set up orchards may be depressed to go ahead and the fruit production in the country would get a set back. Anybody who desires to establish an orchard should hold counsel with such persons who have adequate practical experience in this farming enterprise.

2. Type of Ownership

When an orchard belongs to a college, university, training or research centre, business house, factory *etc.*, evidently the concerned authority becomes the owner of it. The commercial orchard however, if planned to be taken up, proprietors on partnership basis appears to be a better approach than individual ownership and thereby, the shares may be divided among the partners *pro rata* the investments made by them in cash or kind. Large investment is no doubt, possible by joint partnership, besides the fact that the responsibilities may also be shared amongst the partners.

---

[2] There is a saying in West Bengal: Best manure of a crop plant is the dust of the feet of its owner. More frequently it gets, more vigorously it grows and gives rise to high production.

A number of orcharding and farming systems are prevalent across the world. Some of these may be stated as co-operative farming, capitalistic farming, collective farming *etc.* Systems known as contractual farming and contract fruit farming are gaining momentum in India at the present time.

Chapter 3

Selection of Site

3.1 Principles and Objectives

The orchard to be owned by a college, a university, training or research centre *etc.* has to be set up only in the land that is meant for it and evidently, the choice of location is of little importance, except in some special cases.

On the other hand, when an orchard to be run in commercial scale is going to be established, selection of its site is a vitally important step. The owner(s) of such an orchard should note that an orchard may be a highly profitable one if it has been set up in an ideal place, or may be a failure in improper location.

It should be taken into account that a large number of factors become necessary to be considered in selecting the site of a commercial purpose orchard and the pertinent ones may be stated as follows.

1. Fruit Growing Zone

It is advisable that in selecting the site of a commercial orchard, preference should be given to an area which is known to be a fruit growing zone. There are in fact, many advantages if the orchard is set up in a locality where other orchards are already existing, rather than in an area where there is practically no orchard. Among other benefits, the following are of particular mentioning.

(a) Lesser Risk

Orcharding may be a much safer enterprise in such situation and thereby, the risk factors may be less.

(b) Availability of Technical Information

Practical knowledge of fruit culture which has been gained by the fellow orchardists may be obtained as and when required.

(c) Less Recurring Cost

In the existence of a number of orchards at one place, a *co-operative society* may become possible to be formed by the orchardists themselves and by that, the recurring expenditure in functioning of the orchards may be much lessened.

By forming a co-operative society, the equipments which are costly and not always required, *e.g.*, tractors, trailors, power-vans, vehicles, power-tillers, pumping-sets, power-sprayers and dusters, combines and other heavy implements may be purchased and maintained jointly by the members of the society. Similarly, installation of heavy irrigation systems like deep-tubewell, tank,canal *etc.* may also be done on co-operative basis. The co-operative society may be entrusted with other activities as well, notably security system of the orchards and marketing of the harvested produce. A registered co-operative society may even get benefit from the governmental side in purchase of materials at subsidized price and for other purposes.

(d) Less Damage by Animals

It is noted that attack by birds, insects or other animal pests usually becomes lesser in an orchard when other orchards also exits around it. This is due to the fact that in such a situation, the pests tend to distribute somewhat evenly in all the orchards instead of flocking in one and thereby, heavy infestation in a single orchard usually becomes less.

The above beneficent features do not however, necessarily mean that an orchard would count for little, if it is set up in an area where no other orchard exists in the proximity.

In fact, while going to find out a location for an orchard, it is a judicious approach to select a site where other orchards exist in the surroundings. However, in case of unavailability of such a location, it is advisable that an intently desirous person should not reject the site, if it has other merits. Multitudes of highly successful orchards in many counties of the world could be observed which had been set up in areas that were not already spotted with other orchards in the nearness. It is also mentionable that when somebody has set up an orchard in such a location, others would find interest to set up orchards near to it in the near future. Thereby, many orchards would come up in that area in course of time.

2. Physiography

(a) Difficulty in Cultivation

Ploughing, working with heavy implements, spraying, harvesting *etc.* are not easily performed.

(b) Soil Erosion

Erosional loss of soil is high in undulated land, for which roots of trees may be denuded, top soil which is more fertile may be washed away, trees may be lop-sided or lodged.

(c) Difficulty in Irrigation

Surface irrigation is difficult in such a land except by drip- or sprinkler- method.

(d) Water Stagnation

Water logging is a common occurrence in low-lying and sunken areas.

(e) Shading

Some fruit trees always remain shaded by other trees and thus, do not receive adequate sunshine.

(f) Frost Pockets

In cold weather, cold air which is heavier than warm air tends to accumulate in the *sunken pockets* and has difficulty for its easy removal.

(g) Non-uniform Productivity

Productivity of fruit trees varies from place to place in the same orchard and it becomes very low in some parts in such physiography.

3. Market Facility

A business enterprise is like a pipe-line through which water flows. As flowing out of water through one end of the pipe-line is essential for flowing in of water through the other end, so also in commercial fruit growing, earning of profit by disposal, *i.e.*, marketing of the produce harvested from the orchard is essential if heavy expenditure is to be incurred in setting up the orchard and also in its running condition. Thus, it is a very important consideration in selecting the site of a commercial orchard that the produce to be harvested from it would not have any difficulty for quick disposal and at reasonable price.

A number of problems are however, likely to arise regarding efficient disposal of the orchard produce. Those which are more common have been presented in the following.

(a) Non-existence of Market

Absence of any market or a standard market within reasonable distance of the orchard.

(b) Difficulty of Transport

The market may be located at an approachable distance of the orchard but not well-connected with the latter by all-weather motorable road, railway or other easy means of transport.

(c) Costly Transport

The market may be far away from the orchard and thereby, carrying the produce to that long distance market becomes a costly affair.

(d) Malfunctioning of Intermediary

In most parts of India, middlemen are associated with marketing of the horticultural produce. They act in purchasing the produce directly from the growers at the orchard site and then sell it to the wholesale or the retail markets arranging their own transport. For this intermediary activity, they make some earning.

In a healthy marketing structure, not more than 20 per cent of the total profit in raising the produce should actually go to the entire intermediate machinery, which operates from receiving the produce from the growers and delivering it to the hands of the consumers by selling. That is to say, the growers should get a share of 80 per cent of the total profit. But very often it is noticeable that the middlemen seek to derive much more than the 20 per cent share and this is done by them with deceitful measures, some of which are as follows.

The price that is reasonable is not paid by them to the fruit growers on purchase of the produce. Grading of the produce is also not done by them during purchase and only the better quality materials are sorted out rejecting the smaller sized ones. Many of the middlemen take long time to make full payment to the growers. A dirty trick is also done by some middlemen by making unnecessary delay in purchasing the produce from the growers. They deliberately do so with the intention to compel the growers indirectly to sell the produce at very cheap rate. It is a known fact that most of the horticultural commodities perish quickly after harvest under ordinary room temperature condition and thus, become unfit for marketing. In order to get rid of this monetary loss that may take place due to rotting or deterioration of the produce, the growers have no alternative but to accept even very low price offered by the middlemen without bargaining, such that quick disposal of the produce is ensured.

The necessity of intermediaries in marketing of fruits and other agricultural and horticultural produce cannot be practically disregarded in most parts of India under certain conditions. Some of these cold be stated as follows.

(i) When the orchard is far away from a well-established fruit market.

(ii) Transport facility from the orchard to the market is not satisfactory and all-weather motorable road does not exist.

(iii) Orchard produce is much higher, *i.e.*, super-optimum.

(iv) Funding capacity of the orchardist is not adequate.

(v) Harvested produce is of perishable nature.

(vi) Other constraints, that come across directly or indirectly.

Therefore, not only existence of markets but also fair and healthy marketing system near the orchard is of great advantage in orcharding. Evidently, this needs attention in selecting the site, because of the fact that *earning profit in honest way while serving the people should be the motto of a commercial orchard.*

4. Soil Condition

Soil is the *home and burial ground* of the roots of a plant where they grow and are buried when the plant dies, while roots are the *root* of a plant whose duties are anchorage of the plant above ground, taking up water and nutrients from the soil medium and supplying the same to the above-ground parts for production of food and growth.

It is an admitted fact that the soil condition plays a profound influence in successful culture of crop plants. Needless to say that this holds good for fruit crops also. However, it may also be commented upon that undesirable soil condition is not always an uncontrollable impediment in the cropping of fruits. This is because of the fact that many unfavourable characteristics of a given soil are possible to be modified much, so as to suit the requirement of the crop to be grown there by adoption of artificial measures.

For example, the soil organic matter can be significantly enhanced, resulting improvement in soil textural condition by copious application of manure. The fertility status can be heightened by judicious use of fertilizers. Alteration in soil reaction to bring the *p*H to the desired level can also be done without difficulty by proper application of soil-amendments. Among soil factors which is perhaps difficult to alter is its depth. There is also difficulty to get rid of the problems of soil salinity and toxic salts and thereby, render the soil to the effective condition by, artificial means.

In selecting the site of a commercial orchard, evidently the soil factor should be given much importance to ensure its suitability for the fruit crops that are to be grown there. The influence of soil condition in growing fruit crops has been discussed later in this Chapter.

5. Irrigation Facility

Irrigation source such as river, stream, canal, reservoir, large tank, deep-tubewell *etc.* close to the site is undoubtedly a desirable feature in the establishment of a new orchard. In such case, construction of deep or shallow tube-well, tank, well *etc.* may not become necessary to be laid within the orchard land for irrigation purpose. However, if any irrigation source at all exists, the orchardist should also make sure of the following facts.

(a) Right to Use Water

Whether the water that may be available from the source could lawfully be used for irrigating the orchard to be set up there needs to be confirmed.

(b) Water-tax

It any water-tax is necessary to be paid to the concerned authority for using the water, it should not be high and is affordable to the orchardist.

(c) Method of Conveying Water

It should be noted that mere existence of an irrigation source near the site is not enough unless the water from the source is possible to be conveyed inside the orchard land without difficulty.

Such difficulty may arise, if there is any land between the irrigation source and the orchard site and the owner of that land is reluctant to allow the water to pass over his land. Existence of any public road between the irrigation source and the orchard site may also act as a barrier in conveying the water. In the latter case, some growers place water-pipe for conveying the water from other side of the road (Figure 3.1). It should be noted that such an act without permission from the concerned authority is considered illegal. Heaping of sand over the road to make a channel for flowing of the water is also done by many (Figure 3.2), but this is not also a lawful practice.

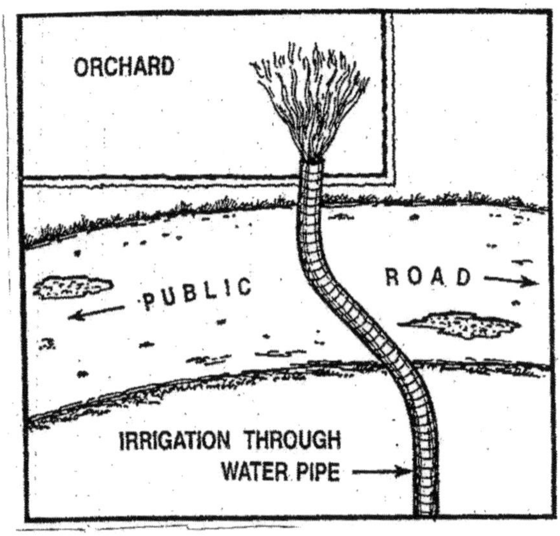

Figure 3.1: Water Pipe Placed over Public Road to Convey Water.

(d) Purity of Water

It should also be ensured that the water from the irrigation source does not contain such substances that are toxic to the crop plants to be grown. It is mentionable in this connection that presence of toxic substances in river water are found in many places in India, particularly when factories are washed with river water and the water carrying the materials is used for irrigation purpose. It is also essential that the water available from the source should not contain salts in so high concentrations as may cause harm to the crop plants, if used to irrigate them. To make sure of this fact, it is advisable that the water should be chemically tested in a reliable laboratory.

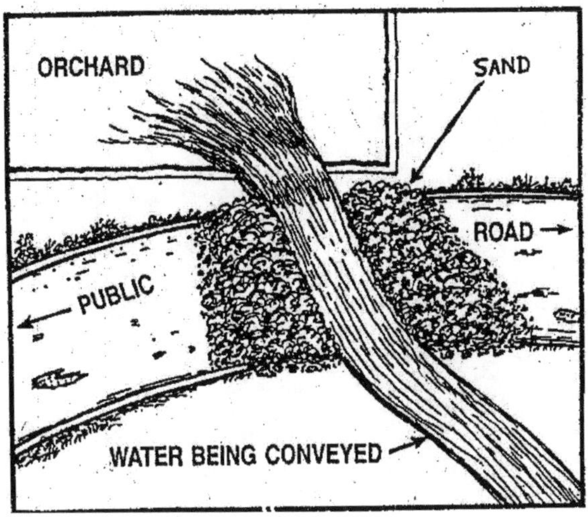

Figure 3.2: Sand Placed Over Road to make Channel to Convey Water.

6. Absence of Flood or Inundation

Under water-logged condition, majority of the fruit plants are suffered, sometimes seriously by showing undesirable symptoms like yellowing of leaves, premature fall of leaves, flowers and fruits in large numbers, drooping of branches and their breaking, stunted growth, physiological rotting of organs, disease infection, low productivity *etc.* Younger trees are generally more easily damaged than the mature ones. Again, trees of some species are more vulnerable and may even be collapsed under submerged condition if last for few days.

Large-scale water submergence is very often a result of flooding by the rivers, although local water due to heavy rain continuing for few days may be cause in some areas. In many parts of India, inundation of large areas persisting for few weeks in rainy-season is a common occurrence in localities, which are situated along the rivers. This is of particular occurrence near the confluences (Figure 3.3), where silting goes on in the rivers and make them more prone to flood along the banks.

Such a site is far from satisfactory in setting up an orchard, except in cases where it is adequately protected against stagnation by effective embankments along the rivers, having suitable arrangements to drain out the water, soil is porous, sub-soil is good and the like.

7. Existence of Processing Factories

Existence of any fruit processing and preservation factory near the orchard site may be of great benefit in regard to disposal of the harvested produce, provided that purchase of the same is regularly done by the factory at reasonable rate from the grower who would set up the orchard.

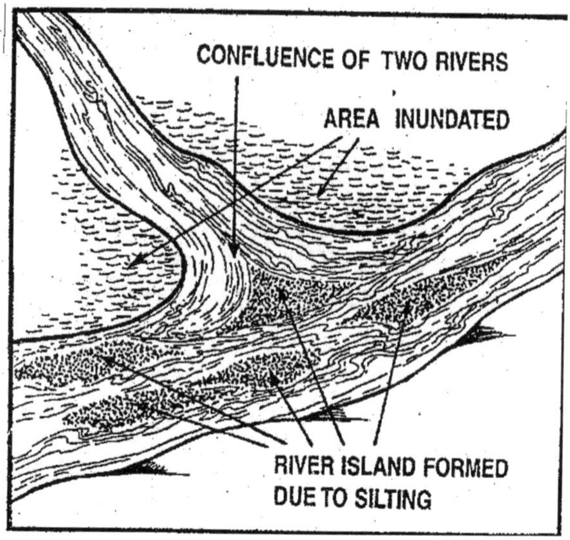

Figure 3.3: Area Inundated Near Confluence of Rivers.

8. Freedom from Storm

Especially in cyclone prone areas, establishment of a commercial orchard may be a risky affair. If the storm is moderate, trees can be saved by planting suitable windbreak rows, but if it comes in the form of a cyclone, windbreak rows usually fail to offer adequate resistance. In such situation, branches of fruit trees may be broken, trees may be felled and even the windbrows may be ravaged.

Some areas are found to be affected by cyclone almost every year. In India, some of the zones which are greatly prone to cyclone may be enlisted as follows:

1. Coastal area in Andhra Pradesh.
2. Saurashtra-Cutchh coastal area in Gujarat.
3. Coastal area in and around Paradeep in Odisha.
4. Coastal area in the Contai sub-division of the Medinipur district in West Bengal.
5. Southern Mumbai coastal area in Maharashtra.
6. Coastal area in Goa.
7. Luni basin in the eastern part of Rajasthan (with occasional cyclone).
8. Northern part of Srinagar in Jammu and Kashmir due to western disturbance, *i.e.*, Mediterranean influence.
9. Brahmaputra valley in Assam.

10. Upper Teesta valley in Sikkim.
11. Western Shimla and north Chhamba area in Himachal Pradesh

9. Availability of Labourers

An important consideration in the selection of an orchard site is availability of labourers. In fruit growing, skilled as well as less skilled labourers become necessary to be engaged depending on the operations to be performed. Hence, it should be ascertained that labourers who are not shirking in habit would be available at the orchard site when required and at standard wage. In the culture of grape or some other fruit crops or for any specialized fruit farming, sophisticated practices are required and accordingly, labourers of even greater skill may be necessary. It should be noted that availability of industrial labourers or masons could not be considered to be a fulfilling condition, because such labourers may be of little practical use for agricultural or horticultural farming.

10. Availability of Manures, Fertilizers *etc.*

Whether manures and unadulterated forms of fertilizers, pesticides, plant growth regulating chemicals *etc.* are available and at standard rate near the site is also a factor of consideration. In such case, the necessity of going to distant places for procurement of these inputs required for cropping would not become necessary and save expenditure to a great extent.

11. Repairing Facility for Machineries

Regular use of implements and machineries is liable to make them damaged, such that repairing of them becomes necessary. Facility for repair of these articles near the orchard is therefore, much helpful and should not be ignored in selecting the site of an orchard.

12. Absence of Industrial Effluents

Existence of a brick-kiln or factory near the orchard site that billows smoke or discharge dust particles is no doubt, an undesirable feature. Exposure to smoke causes great harm to the orchard trees or the fruits borne by them, an example of which may be cited as the physiological disorder, known as *black-tip* occurring in the fruits of mango under north Indian conditions. The symptoms appear as black patch at the tip of the fruits which later discolour with necrosis extending inside the pulp and formation of fissures in the affected parts, giving rise to oozing out of a brown, gummy liquid (Figure 3.4).

The smoke from factories may contain hydrocarbon gases and thereby, this gas is liable to cause shedding of leaves, flowers or fruits, premature ripening of the fruits or untimely flowering of the trees and many other abnormalities. Existence of thermal power station near an orchard creates a problem in that, the coal-ashes discharged from it blow away and provide thick layers on leaves and flowers and thus, hamper photosynthesis, transpiration and pollination. But electrostatic precipitators are used by many stations which collect the ash. (However, cinders, if available from the thermal power stations are valuable materials for partial filling of the pits that are made for planting).

Figure 3.4: Black-tip Disorder of Mango.

13. Absence of Electric Lines

A land over which electric lines run is considered unsafe in setting up an orchard. This is because, the fruit trees being tall may come close to or in direct contact with the electric wires carrying high voltage and result in serious accident. High wind may also cause tearing away of the electric wires and felling them on the ground or even felling of the poles on the ground. These may even result in electrocution of the workers in the orchard.

Setting up an orchard just below a ropeway has also possible risk.

14. Street Light

If the orchard is situated along a road, which remains brightly illuminated throughout every night, sometimes it is seen that the orchard trees, especially those which are adjacent to the roads produce flowers out of season, or fail to enter into dormancy in normal time if deciduous in nature. These phenomena may occur due to alteration of the normal *photo-dark periodic cycle,* which is effected by the street lights. The day-neutral plants though remain unaffected by such light exposure, the *photo dark periodically sensitive* plants are influenced by this factor and show such effect. For this reason, the horticultural scientists in fact, hesitate to recommend such a site for commercial orcharding. (Strictly day-neutral plants are however, not many).

15. Social Factor

A site in a locality inhabited by pilferers and anti-social elements is by no means, satisfactory to establish a commercial orchard, even though it is ideal in other respects. Many orchardists have also the experience that the local people have forced them to provide jobs in the orchard but on getting the jobs, they show unwillingness to do any type of work. When asked the reason, they threaten the orchard owner and even resort to subversive activities than giving any satisfactory answer. Even some so-called, dishonest local leaders give indulgence to these knaves for petty gain. The orchardist comes to realize his mistake in setting up the orchard in such a dirty situation only to invite troubles everyday and cosiders of selling out the orchard at much lower price or stops all works in the orchard.

Besides the above factors, there are also many other considerations which are necessary to be kept in view in selecting the site of a commercial orchard. Selecting an orchard site without heedful attention to these aspects will only erode money and peace of mind of the orchardist.

3.2 Size of Orchard

An orchard may be very small to very large according to choice or the land that is available. But for a commercially run orchard, area of it is an important consideration. If the area is too small, use of heavy implements and mechanization in cropping have less scope and hence, extent of return is lesser than an orchard of large area where *mechanized farming* has greater opportunity. On the other hand, a very large orchard may not prove its worth if it is not manageable. Thus, for a commercial orchard in general, larger area is always preferable provided the management is efficient. In any case, it can be said that a commercial orchard of less than a hectare area of land is not befitting in most cases.

3.3 Orchard in Undesirable Site

Although the conditions enumerated above appear to be highly suitable to set up new orchards, it should be noted that establishment of orchards cannot be ruled out if such conditions are not available by any means.

In India, about 63.85 million hectares are considered as wasteland and this comprising 20.17 percent of the geographical area grow very little food crops. The waste lands in the country are usually categoriesed as gullied and/or, ravenous land, upland with or without scrub, waterlogged and marshy land, saline or alkaline land, grazing land *etc.* Many of such types of land in different parts of India are being attempted to reclaim so as to transform these lands for raising food crops. It may be mentioned that plantation of fruit crops and especially those which are hardy and recognized as less utilized (minor) fruit plants might be suitable ones in such lands after reclamation.

Drought is another hazard in many parts of India which stands in the way of crop cultivation in an efficient way. Drought is dryness of soil due to lack of rain or extremely low rain and this may be due to meteorological, agricultural or hydrological. Even large-scale plantation of drought resistant or tolerant fruit plants should be judiciously considered in such area for cropping.

Saline and alkali soils pose great threat in crop cultivation. Much area under such soils are found in Haryana, Punjab, Uttar Pradesh, Bihar, Andhra Pradesh, Karnataka and in some other States. The *p*H and Exchangeable sodium percentage of saline soils are less than 8.9 and 15 but in alkaline soils, these values are higher than that.

Although all fruit crops are not suitable to be grown in these types of soils, there are some which are moderately or highly salt tolerant. Among highly salt tolerant fruit crops, mention may be made to coconut, karonda, loquat, date-palm, wild date-palm *etc.* Some of the moderately salt tolerant fruit crops are cashew, custard apple, atemoya, bullock's heart, akee (blighia), white sapote, star apple, diospyros, pineapple guava, fig, mango, sapota, passion fruit, guava, rose-apple, cattley guava, Suriname cherry, Barbados cherry *etc.* These fruit crops may be grown. Salt content in these soils may however, be reduced to some extent in such soil by mechanical removal of salt deposited on surface and heavy watering at frequent intervals. To grow fruit plants in soils of higher salt concentration, *nitrate fertilizers* should be used rather than the ammoniacal ones in respect of nitrogen fertilization.

Many orchardists are not found to be interested in soil of such type which is formed on drying of river bed. This is not correct. In fact, such type of soil is highly fertile and are ideally suitable for fruit growing if not inundated by change of the course of river. (In some parts of north India, people say that the course of the Ganga river changes after every fifty years and renders vast land free for cultivation of crops).

Chapter 4

Selection of Fruit Crops

4.1 Principles and Objectives

If an orchard is set up for experimental purpose or for other objectives, any type of fruit crop(s) may be grown according to the *choice of the orchardist.* But in a commercial orchard, *profit from it is the main or only consideration* and hence, such fruit crops(s) should be preferred that would give rise to profit to the orchardist. To establish a commercial purpose orchard, which is the primary objective of the present text, the orchardist has in fact, to pay attention on a number of factors. It may be brought to notice that among other considerations, the orchardist should pay much greater attention on the following aspects in particular.

1. Climatic Suitability of the Species of Fruit Crops

As stated in Chapter 1, fruit crops of different species have different types of climatic requirements for their healthy growth and satisfactory production. In accordance with, they are grouped as tropical, subtropical and temperate fruit crops. It may be a ridiculous approach if a tropical fruit crop is attempted to grow in a temperate climate or *vice versa* in a commercial orchard in which, profit remains to be the objective and not experimentation. It may however, be mentioned in this connection that some species of fruit crops are not very exacting in their climatic adaptability. For example, mango, guava, mandarin orange, jackfruit *etc.* are some examples that grow well in both tropics and subtropics while fruits like pear, peach, plum, strawberry are some examples that grow in subtropical and temperate climates however, depending on varieties.

It may be mentioned in this connection that requirement of specific type of climatic condition is felt by the fruit crops at some particular stage(s) of their growth, flowering and fruiting time. When such conditions are not obtained in those stage(s), their growth and productivity are hampered. For example, in a temperate fruit like apple, a chilling exposure of 4°C or less for 2 months or more is almost essential to break dormancy of buds, so that growth of shoots start and flowers are borne by them in spring.

For subtropical fruits, although exposure to freezing or to very low winter temperature is not required, distinct *chilling in winter* is in fact, necessary for many of them to produce fruits of superior quality. This happens in many fruits of tropical and subtropical climate and above all, a classic example is guava, sugar formation in which fruit takes place at higher rate on receiving low temperature exposure. Fruits are found to be sweeter within a day or two of sudden drop down of temperature when it occurs in winter. Pomegranate could be cited as another example of this group. In fig, the cooler night temperature is required for increased sugar formation in the fruits. The colour development is also found to be related to low winter temperature in some fruits.

Instead of low temperature, high temperature in summer is also considered beneficial for some fruits, especially for better ripening of them and date is a suitable example of this group. Shaddock and some early varieties of pear also seem to require warmer atmosphere to develop fruits of better quality. High summer temperature is however, known to cause abnormalities as well in some fruits. A notable example is the symptom of granulation of juice which is observed in mandarin orange, grapefruit and other citrus fruits. Although improper rootstock, delayed harvest *etc.* are responsible for this abnormality, high summer temperature is also considered to be a cause. In walnut, formation of *blank nut* and sun-burning of many fruits in northern India are known to take place by high summer temperature.

Evidently, high summer temperature is more harmful, if it is accompanied by high wind. Such condition in northern India is known as *looh* and is highly detrimental for abnormality in many fruits. Loquat is an example, fruits of which become smaller-sized and ripen improperly in such condition. In litchi, hot and dry atmosphere in summer is by far, the most important limiting factor for its widespread culture in India.

Damaging effect of high wind is also known in many fruit trees, even in absence of high temperature. In banana, high wind is a destructive climatic factor that causes drying of the leaves, breakage of pseudostem, damage to roots *etc.* Wind velocity above 60 kilometers per hour is highly harmful to the banana plants and the velocity above 95 kilometers per hour is in fact, liable to uproot the plants.

High wind is also a threat for tilting, lodging or uprooting of trees and especially the shallow-rooted fruit plants like papaya, custard apple, sapota, fig, phalsa, pineapple *etc.* High wind, if accompanied by dry weather is again, a serious bottleneck in fruit culture. Such atmosphere is responsible for severe moisture stress to the fruit trees due to high rate of evaporation from soil surface and evapo-transpiration from the leaves.

Rainfall is beyond doubt, another major climatic factor that needs critical consideration. As regard this factor, it is an important point to note that distribution of rainfall throughout the year has a greater impact on vigour of a crop plant and its productive capacity than the total amount of rainfall actually received by it in a year. Tropical fruit crops always prefer high rainfall areas and their performance is better, especially when grow in the coastal tropical parts. Requirement of rainfall is however, relatively less for many subtropical and temperate fruit crops. It is also of interest to note that some subtropical fruit plants change their growth pattern and flowering when grown in tropical climate. Thus, grape, pomegranate and some other fruit plants which are deciduous in subtropical climate are found to be much vigorous and remain evergreen, or defoliate partly when grown in tropical situation having high rainfall and mild winter.

Fruitfulness is greatly influenced in many fruit plants by high rainfall such that, it is favourable for some but harmful for other fruit plants. Taking the case of fruit set, it is seen that high rainfall and high relative humidity during flowering time hamper pollination and thereby, results in shedding of flowers in fruit trees like mango, jamun date-palm, cashew *etc.* On the other hand, high humid atmosphere during flowering time favours pollination in some fruit trees and notably in custard apple.

High relative humidity during development of the fruits is conductive for some fruit plants and especially in litchi, where dry atmosphere causes splitting of the fruits. On the other hand, fruits like papaya, guava, sweet orange, fig *etc.* become qualitatively poor and taste insipid when develop and ripen in high humid atmosphere. Highly humid atmosphere is also a causative factor for higher incidence of diseases like bacterial canker in sour lime and virus infection in papaya. The physiological disorder in the fruit, viz., mangosteen, which is known as *gamboge* and is characterized by gummy exudation from fruits results due to high rainfall that occurs during development of fruits.

The above are only few examples of good or bad effects of the climatic factor on fruit culture. Hundreds of examples may in fact, be cited to show clearly the definite role of climatic components like temperature, rainfall, relative humidity, light intensity, photo-dark periodic and thermo-periodic cycles, wind velocity *etc.* in controlling growth, flowering, fruit setting and productiveness of fruits, in respect of yield performance and fruit quality in a vast majority of fruit crops.

Consequently, the orchardist must have a sound knowledge on the effect of climatic factors in controlling the quantitative and qualitative productivity of the specific fruit crops. If he lacks this knowledge, he should gain it from persons who are highly experienced in fruit culture or, consult authentic literature for that area. In consideration to this aspect, he should judge what fruit crops would be sustainable to be grown in the site which he has obtained to set up the orchard, with the idea that high profit is expected to come from the orchard. This is of great significance in a commercial orchard, which aims at securing profit.

2. Climatic Suitability of Varieties, Hybrids etc.

Some particular variety or varieties or hybrids of many species of fruit crops are found to have some specific type of climatic requirements. That is, a variety of a fruit crop makes better performance in one area, while another variety does better in a different zone. For example, in mango, the langra variety is at its best in Banaras region, Chausa in Lucknow-Malihabad region, Dusheri in north India, Himsagar in West Bengal, Baneshan in Andhra Pradesh and the Alphonso variety, strictly in Ratnagiri and Konkan regions in the west coast of India. Similarly, Safeda variety of guava is known to be at its best when grown in the Allahabad area of U.P., Mosambi variety of sweet orange in Pune and Santra vairiety of mandarin orange in Nagpur region of Maharashta, Ambri variety of apple in Kashmir, Kew group of pineapple in the north-eastern States, Anab-e-shahi variety of grape in the Hyderabad region, early varieties grape in the Punjab, Haryana area and so on. This is because of the fact that the specific requirements of some varieties or hybrids for better performance are met with in some specific eco-environmental conditions.

In setting up of an orchard for commercial purpose, adequate emphasis should be given in regard to choice of varieties of any species of fruit crop which is to be grown in the orchard. Precisely speaking, the orhardist should only plant that variety(ies) or hybrid(s) of a fruit plant, which is known to have proven performance in that agro-climatic situation. If this is not in his knowledge, he should consult fellow orcharadists in the locality or a specialist and be confirmed of the fact.

In the following has been presented a list of a few varieties of some important fruit crops that are known to have greater suitability in different parts of India but it should be noted that there are many other examples apart from these.

(a) Mango

(i) North India

Dusheri, Langra, Chausa, Rataul, Bombai green, Zafran, Fazli, Malihabad safeda *etc.*

(ii) Bihar

Sukul, Sepia, Dudhia, Mithawa, Hathijhool, Bathua, Zardalu *etc.*

(iii) West Bengal

Himsagar, Safdar pasand, Rani pasand, Shah pasand, Bira, Bhuto Bombai, Golapkhas, Kisenbhog, Jhumko fazli, Sorikhas, Peyaraphuli *etc.*

(iv) Andhra Pradesh

Baneshan, Imam pasand, Banganpalle, Swarnarekha, Neelum, Panchadarakalasa, Cherukurasam, Rumani *etc.*

(v) Karnataka

Totapuri, Mundappa, Neelum, Ratna *etc.*

(vi) Tamil Nadu

Neelum, Mulgoa, Rumani, Jehangir, Totapuri *etc.*

(vii) Kerala

Mulgoa, Neelum, Rumani, Mundappa *etc.*

(viii) Maharashtra

Mankurad, Pairi, Alphonso, Kesar *etc.*

(ix) Konkan

Alphonso, Fernandin, Pairi, Mankurad *etc.*

(x) Gujarat

Rajapuri, Vanraj, Jamadar, Pairi *etc.*

(xi) Rajasthan

Ras Bhandar.

(b) Banana

(i) Kerala

Kunnan types, Safed velchi, Poovan, Chen kadali, Nendran *etc.*

(ii) Tamil Nadu

Poovan, Harichhal, Rasthali, Safed velchi, Dwarf cavendish *etc.*

(iii) Maharashtra

Dwarf cavendish, Safed velchi, Harichhal *etc.*

(iv) West Bengal

Rasthali, Amrit sagar, Poovan, Anupam, Malbhog, Dudh sagar, Giant Governor Anupam, Dwarf cavendish *etc.*

(c) Sweet Orange

(i) Punjab

Malta, Pineapple, Hamlin, Valencia, Malta blood red *etc.*

(ii) Maharashtra

Mosambi.

(iii) Andhra Pradesh

Sathgudi, Batavian *etc.*

(iv) Tamil Nadu

Sathgudi

(d) Mandarin Orange

(i) Maharashtra

Nagpur santra.

(ii) West Bengal

(Himalayan region) Darjeeling (syn. Sikkim).

(iii) North-east India

Khasi, Sylhet, Cleopatra, Dancy *etc.*

(iv) Karnataka

Coorg.

(e) Kinow Tangor

N.B.: Kinow is an orange hybrid, *i.e.*, tangor (tangerine × orange) and has been produced by *hybridization* between King orange and Willow leaf mandarin orange. It behaves like mandarin orange but it is semi-loose-skinned. (It is wrongly termed as know mandarin)

(i) Punjab, Haryana and north India

Kinow tangors of different sizes and rind looseness.

(f) Guava

(i) North India

Allahabad safeda, Lucknow- 49, Chittidar, Banarasi, Apple guava *etc.*

(ii) Bihar

Hari Jha, Hafsi, Seedless *etc.*

(iii) West Bengal

Baruipur, Hafsi, Hari Jha, Chittidar *etc.*

(iv) Maharashtra

Nasik, Dharwar, Sind, Dholka, Red banana *etc.*

(g) Grape

(i) North India and Punjab

Perlettle, Pusa seedless, Delight, Gold, Khalili, Himrod, Dakh, Early muscat *etc.*

(ii) Andhra Pradesh

Anab-e-shahi, Bangalore blue, Selection-7, Regina, Pusa navrang *etc.*

(iii) Maharashtra

Bhokhri, Cheema sahebi, Kali shahebi, Kandhari, Thompson seedless, Tas-e-Ganesh, Sonaka *etc.*

(iv) Karnataka, Hyderabad, Bangalore

Anab-e-shahi, Bangalore blue, Arka hans *etc.*

(v) Tamil Nadu

Bhokri, Gulabi, Thompson seedless, Pakchai drakhshai *etc.*

(vi) Himachal Pradesh (Kinnaur area)

Kandhari, Hussaini *etc.*

(h) Apple

(i) Himachal Pradesh

Golden delicious, King of Pippins, Starking delicious, Mc Intosh, Baldwin, Red gold *etc.*

(ii) Jammu and Kashmir

Ambri, Cox's orange pippin, Red delicious, Golden delicious, Jonathan, Benoni, Irish peach *etc.*

(iii) Punjab Hills

Beverley hills, Tropical beauty, Rome beauty, Winter banana, Namna *etc.*

(iv) Uttar Pradesh (Hills)

Fanny, Jonathan, Benoni, Red delicious, Golden delicious *etc.*

(v) South India (Nilgiri Hills)

Rome beauty, Irishi peach, Zouches, Carrington *etc.*

(i) Pear

High attitude: Bertlett, Conference, Bagu gosha, Flemish beauty, Anjou *etc.*

Sub-montane area: China, Keiffer, Gola, Patharnakh *etc.*

(j) Peach

High attitude: Elberta, J.H. Hale, Prince of Wales *etc.*

Sub-montane area: Sharbati, Safeda, Early amber, Saharanpur-6 *etc.*

3. Varieties of Higher Productivity and Better Quality

Needless to say that in a commercial orchard, such varieties of fruit crops should be grown which fetch high economic return. Accordingly, those varieties whose *productiveness is higher*, evidently needs emphasis. However, it should also be remembered that *quality is a very important criterion of the fruits to earn profit* from a commercial orchard. Thus, a few fruits of some choice varieties of mango like Alphonso, Banarasi Langra, Samar bashist chowsa, Mallika, Amrapalee, Arka puneet, Himsagar *etc.* would fetch much higher return than a basketful of inferior varieties.

Some varieties of fruit crops are found to be less fruitful with pollens of flowers of the same varieties. Hence, if these varieties are grown, such varieties whose pollens would be effective to fertilize flowers of self-unfruitful, or less fruitful varieties should be grown together, or at nearness to them.

Therefore, in regard to choice of varieties, hybrids *etc.*, great attention should be paid to secure those which give rise to higher production and better quality of fruits.

4. Varieties having different Harvest Time

In many species of fruit crops, fruits of all the varieties do not become ready for harvest at the same time. That is, in some varieties, the fruits attain harvest-maturity earlier while in some varieties later. Accordingly, the varieties are grouped as early and late and for some fruits, mid-season maturing varieties also exist when the harvest time is intermediate between the early and the late ones.

If only one variety of a species of fruit crop is grown in the orchard, fruits of all the trees are needed to be harvested at same time. Such an orchard has a risk in that, the large number of fruits that are harvested at one time may have practical difficulty in marketing. Besides, the market price may also become much abated in such case due to greater supply of the produce than the demand that is felt. On the other hand, if different varieties are grown having different time of harvest, fruits will be obtained over a longer period. By that, not only the possible risk factor that is feared in growing only one variety could be avoided but the income from the orchard would also be ensured over a longer period.

Therefore, it is advisable that in a commercial orchard which primarily aims at profit, *mixed varietal cropping* is considered desirable, In such case, judicious selection of varieties that are,

(i) *adaptable* to the locality,

(ii) yield the produce at different times, and

(iii) have demand in the market or have no obstacle in disposal of the produce should be critically considered.

5. Varieties having Less Risk

In some species of fruit crops, there are varieties that may produce excellent fruits but these varieties may also have some problems with them, such that their cropping may involve risk. One such risk factor is *biennial* or *irregular bearing*, which is characterized by failure to produce flowers in optimum number in every year.

The above malady is known to be inherent or more acute in some varieties of mango, apple and some other fruit crops, among which, mango is perhaps the worst sufferer. Unfortunately, the intensity of the problem is met with in mango in some choice varieties and perhaps the Langra variety tops the list. By growing these varieties, evidently the grower has to suffer an economic loss when a tree is in "off-year" condition and the loss is extreme when no flower is produced in the "off-year". On the other hand, fruits of these varieties having high preference of consumers fetch high return in the "on-year".

Thus, as a compromising measure, it is advisable that in a commercial orchard, not only the biennial or irregular bearing varieties should be grown, despite the fact that they are choiceable ones, but should also be planted with some *regular* bearing varieties, *i.e.*, such varieties in which the bearing problem does not exist or exists only mildly. Some of the *regular* or *moderately regular* bearing varieties of mango grown in different parts of India may be named as Dusheri, Fazli, Amrapali, Mallika, Dalma, Brindabani, Kamia, Kalapady, Totapuri red small, Bangalora, Romani Baramasi, Himsagar *etc.* In apple, some of the regular bearing varieties are Mc Intosh, Jonathan, Rome beauty, Spartan, Ben Davis, Stayman, Winesap, Cortland, *etc.*

Varietal difference on susceptibility, tolerance, resistance and even immunity to many diseases is apparent in many species of fruits. A few examples are stated in the following. In case of banana, the varieties, Rasthali, Martaman, Monthan, Alpan, Virupakshi *etc.* are found to be susceptible to the Panama disease, while Poovan, Moongill, Vamankelli *etc.* are resistant and the Basrai is perhaps immune to the disease. The most serious fungal disease in guava, *i.e.*, wilt takes the toll of a large number of trees every year in India for some varieties and notably the famous one *viz.*, Allahabad safeda, while the varieties, Banarasi, Nasik, Dholka, Supreme *etc.* show resistance to this disease. Mention may be made to the fungal disease, leaf spot in sapota to which it has been experimentally proved that the varieties, Kalipatti and C0-2 are tolerant, the variety, Cricket ball is resistant and the Kalkatia round variety is highly susceptible. The variety, Jonathan in apple is known to have high susceptibility to the fungal disease, powdery mildew.

In the selection of varieties of fruit crops, it is accordingly advisable not to take up those which are highly vulnerable to diseases or abnormalities as may jeopardize the productiveness.

6. Growing Several Species of Fruit Crops

Growing more than one species of fruit crops in a commercial orchard has a number of benefits in many cases. Perhaps the greatest advantage in such cropping lies in the fact that the possible risk that may arise by growing a single species of fruit crop is avoided. That is, if only one species of a fruit crop is grown in the entire orchard and if the crop is lost in some year due to natural calamity, attack by pests or diseases, or due to other reasons, the orchardist has to suffer a total loss in that year. But this loss is possible to be compensated from the production that could be obtained from fruit crops of other species, if grown in the orchard. Again, the harvest time of the fruits under different species being different, earning from the orchard is also possible at more than one season.

If fruit farming is done by growing fruit crops of many species in the same orchard, it is advisable for the orchardist to keep in mind the following facts.

(i) The fruit crop belonging to a species should be grown exclusively in a *separate block*. That is, they should not be inter-planted with other species, except when *filler trees* or *shrubs* are considered to be planted.

(ii) Dwarf species of a fruit crop should not be *shaded* by the tall fruit trees.

(iii) As far as possible, planting should be done in consideration to *east-west* direction (Figure 4.1)

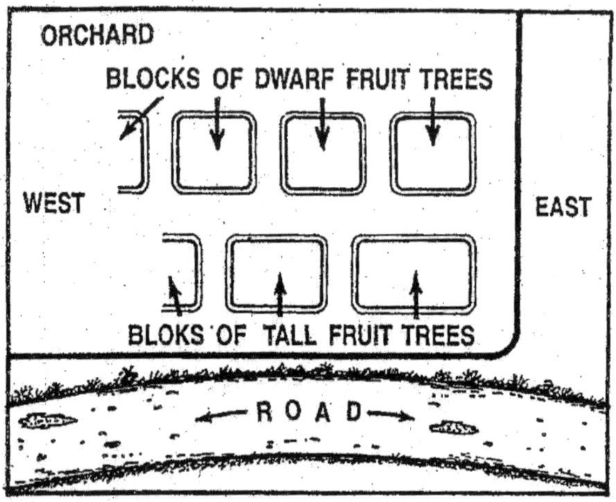

Figure 4.1: Blocks of Fruit Trees Allocated to Avoid Shade of Tall Trees over Dwarf Ones

7. Edaphic Suitability of Fruit Crops

Whether the fruit crops to be grown are adaptable to the soil condition of the land which has been obtained to set up the orchard evidently needs to be given adequate attention. However, as stated earlier, some of the conditions of soil though not all, are possible to be improved to some extent to suit the requirements of the fruit crops. Nevertheless, in the selection of fruit crops to be grown, the soil condition of the land should be adequately kept in view. The important soil factors that are necessary to be considered have been stated in the following.

The depth of the soil has its importance in that, fruit crops belonging to different species have different penetrating depth of their roots. Thus, fruit trees like mango, jackfruit, some citruses, loquat, pear, plum, walnut *etc.* forage their roots to a relatively great depth and hence, a depth of 1.5-2 metres appears to be necessary for proper growth of them.

For other fruit crops like banana, papaya, phalsa, fig, pomegrantate, grape, apple *etc.*, a depth of upto 90 cm seems to be optimum. Again, there are fruit trees like sapota, custard apple *etc.* which may grow well in soil of upto 75 cm depth, while for fruits like pineapple, strawberry and cape-gooseberry, soil depth of 40-45 cm may be considered enough.

So far as texture of the soil is concerned, it may be stated that most of the fruit crops prefer *sandy loam* or *loamy soils.* Heavy soil is in general, unsuitable for fruit crops because of poor drainage. However, some fruit trees like fig. date, litchi, pear, some citruses are found to grow well in such soil, provided the water management is done efficiently.

The coarse-textured soil has a drawback in that, the surface layer is dried up more readily and hence, frequent irrigations are to be given in such soil. Accordingly, such soil is suitable in area where rainfall is prolonged as in the Khasi hills and the north-eastern hilly region where good quality manadarin oranges are grown in sandy soil. In regard structure, the soil with open, crumb or granular structure could be regarded to be much suitable for majority of the fruit crops.

Perhaps the most important soil factor for successful culture of fruit crops is *drainage*. Some fruit trees are particularly sensitive to poor drainage among which, jackfruit, oranges, avocado, papaya, limes and lemons are notable examples.

The performance of many fruit crop is better when they are grown in a specific range of soil *p*H. For example, in pineapple cultivation, a soil *p*H of 5 or even low is considered desirable, particularly as iron uptake is ensured by the plants in acid soil, which is necessary for this crop. Likewise in strawberry, 4.6-6.4 *p*H is considered optimum. For peach, soil would be suitable having a *p*H of 5.8-6.8 and mandarin orange is found to grow satisfactorily in soil *p*H of 4.5-5.5 in many parts of India. Guava trees can grow over wide range of soil *p*H, but in alkaline soil, the chance of attack by the serious soil-borne fungal disease, guava wilt is greater.

Slightly alkaline soil is considered helpful in banana culture due to the fact that wilt diseases appears in the plant in acid soil. Some other fruit trees are also known to prefer alkaline soil, or the tolerance capacity to soil alkalinity could be said to be greater in them. An example is litchi cultivation in north Bihar, where perhaps every grower applies lime in soil and in that area, the litchi trees are found to tolerate even 30 per cent lime. However, whether litchi trees really need liming is not very clear, because in many parts of the world, it grows quite satisfactorily in neutral or even acidic soils. Date-palm is mentionable, which is able to tolerate 2-3 per cent of soil alkali. High amount of alkali in the soil is found to be withstood by figure also, but it needs mentioning in this connection that even very low level of *black alkali* (non-saline alkali soil due to high sodium carbonate) proves injurious effect to this tree. A fruit tree, custard apple could no doubt, be said to be a classical one which grows and gives fruit well in very inferior type of soil and tolerates as high as 50 per cent lime in the soil.

It goes without saying that soil fertility is a very important consideration in the culture of any crop and overthly the, fruit culture is no exception. Nevertheless, it may be pointed out that there are some fruit trees, notably, ber, phalsa, custard apple, pomegranate, guava *etc.*, which have the potentiality to grow and produce well in soils that are relatively less fertile. However, extremely poor soil is almost unsuitable for production of fruit crops, but it is possible to enhance its fertility by efficient use of manures and fertilizers.

A high level of soil salinity is definitely injurious to fruit crops but the degree of tolerance is known to differ amongst the species. For example, fruit crops like peach, avocado, strawberry *etc.* can be named, which are extremely sensitive to soil salinity and cannot endure more than 3.5 mmohs/cm. Other fruit trees *viz.,* apple, pear, plum, pomegranate *etc.* are there that can tolerate soil salinity of upto 6 or 6.5 mmohs/cm, while tolerance level may even go upto 8-8.5 mmohs/cm for fruit trees like date, guava, grape *etc.*

Besides the above, other soil factors also play part and these should no doubt, be taken into consideration in regard to selection of fruit crops to be grown in a new orchard.

8. Market Demand

The importance of market facility in selecting the site of a commercial orchard has been indicated earlier. In the same way, the demand of the harvested produce in the market needs emphasis in the selection of fruit crops to be grown in the orchard. Marketing of the fruits may be done to the nearby markets or to those located at distant places. As a commercial orchard aims at profit, the orchardist must know where the produce could be marketed for higher return. Evidently, if higher earning is ensured from the local markets, he should sell it there. The requirement of the fruit in the market where the major part of the produce is to be sold should therefore, be properly assessed and the fruit crops to be grown should also be selected accordingly.

It is also pointed out in this connection that not only species of fruit crops but even some specific *variety* or *varieties* within a species have a greater demand in many markets. Interestingly, it is observed that although many of such choice varieties of a fruit have a higher demand in the market of a locality, the demand may be felt much lesser in other markets and sometimes even in the same locality.

9. Supply of Produce to the Processing Factories

Selection of the fruit crops that are to be grown in a commercial orchard may depend on another factor as well, that is, whether any fruit processing factory exists near the orchard site and if so, whether the harvested produce of the orchard could be sold to the factory. In such case, the orchardist should also confirm the following facts before deciding the fruit crops to be grown.

(i) Whether the produce of the orchard that is to be grown would be purchased by the factory.

(ii) How much quantity of the produce would actually be taken by the factory from him.

(iii) At what price, the produce would be purchased by the factory and whether, that price is reasonable so as to ensure profit by the orchardist.

(iv) Wether the factory would purchase the produce from him on a long-term basis, or for one or a few years only.

Only when the orchardist gets satisfactory answers to the above questions, he would go for selecting such type of fruit crops for growing in his orchard for the above purpose.

Thus, to meet the requirement for making fruit products like jelly or jam, those fruits which are rich in pectin of high jelly-grade, e.g., pear, guava, pineapple, plum, loquat, strawberry, Japanese plum, peach, bael, ber, karonda, cape-gooseberry etc. are suitable materials. Similarly, in making marmalade, citrus fruits such as jambhiri, karna, galgal, calamondin, Nepalese oblong, ponderosa lemon etc. appear to be appropriate materials. In the same way, for making squashes, juices and beverages, fruits like mandarin orange, pineapple, grape, litchi, pomegranate, jamun, juicy mangoes, phalsa, watermelon etc. are considered suitable material. For conning purpose, fruits like pineapple, mango, litchi, peack, pear, bael, ber, wax-apple, rose-apple, palmyra seeds etc. are much preferred under Indian condition. It is also pointed out in this connection that for the purpose of processing, there may be specific requirement not only for the fruits of any particular species, but even for particular varieties, strains, types *etc.* within that species.

For example, in the canning of pineapple, fruits of the *Kew* group are preferred than those of the *Queen* group. Although, fruits of *Queen* are sweeter and have higher fragrance than those of the former group of fruits, these are usually not preferred in the canning factories. This is because, the *eyes* which are present on the surface of the fruits are deeply set in the Kew fruits and hence, trimming of the fruits, *i.e.*, removal of the skin by machine (as is done in the factories) involves removal of a considerable part of the pulp, in order to make it absolutely free from any scar of eye. In other words, much of the pulp is actually wasted. On the other hand, in the Kew fruits, the eyes are shallow and not deep set, for which, trimming with removal of eyes is done in machine with lesser loss of pulp.

For canning of other fruits also, there is a preference for particular varieties. For example, in canning of mango, fruits of the Alphonso variety are considered highly suitable as the pulp is not soft and has a characteristic flavour, which is retained in the canned product also. In canning pears, Bartlett variety has a preference and in canning peaches, fruits of the varieties, Elberta, J.H. Hale, Fortuna, Carolyn, Cortex Dixon 1, Halford, Vivian, Crawford's early *etc.* are desired. For grape canning in India, berries of the Thompson seedless variety are much in use. In the canning of strawberry, fruits of some varieties like Talisman, Royal sovereign, Cambridge, vigour, Cambridge favourite are considered superior. If litchis are canned, there is a choice for fruits of some varieties, *viz.*, Muzaffarpur, Bombai, Rose scented, Purbi, Early bedana *etc.*

Again, in the preparation of squash from mandarin orange, fruits of the variety, Nagpur santra are largely used in India for some qualities like abundant juice with excellent flavor, saffron yellow colour of the juice, total solids of 11-13 per cent and acidity of 0.5 per cent in the juice, less seed content in the fruits *etc.* Fruit juice of apple varieties, *viz.*, Freedom, Majda *etc.* also gains prominence in making bottled juice, or cider (= mildly fermented juice of apple). Mango fruits of some juicy varieties like Raspunia, Mithawa Gazipur, Sarbati, Mithawa sundershah, Gaurjit, Gilas, Yusuf pasand, Hardal Aziz *etc.* have greater suitability to make bottled juice or squash. In making grape juice, berries of the varieties, Eearly Muscat, Black champa, Bangalore, blue, Bharat early Champion *etc.* are used in India. To make squash from litchi, juice of some varieties, *viz.*, Kalkatia, Rose scented, Early bedana *etc.* are considered desirable.

Preference is also there for the Allahabad safeda variety of guava in making jelly. To make strawberry jam, the varieties, Cambridge vigour, Talisman and Red gauntlet are considered superior. In raspberry, fruits of the variety, Person's red are known to have greater demand to make jam.

In making pickles from olives, the varieties, Kalamon, Nocellara eatnea, Tonda Iblea, Picholine *etc.* and from aonla, the varieties, Banarasi, Chakaiya *etc.* are preferred in the processing industry. In the Maldah district of West Bengal, mango leather is produced using fruits of the variety, Fazli almost exclusively.

Only few examples have been cited in the foregoing pages. Besides those, there are plentiful examples of the varieties of fruits under different species that have greater suitableness in making specific type of preserved products. As the fruit processing and preservation factories have a demand of the fruits belonging to such varieties, the orchardist should resort to crop with those varieties. Lastly however, it is also pointed out that instead of supplying the fruits to other factories, the orchardist may even take the step to set up such a factory himself or with partners which would make utilization of his orchard produce as the raw materials.

Even prematurely dropped fruits could be made use of in making pickles or other fruit products. It may be mentioned in this connection that some species of lesser known fruit crops (minor fruit crops) are grown in inferior land in some parts of India for land utilization and to supply the produce to the fruit processing factories. Cattley guava may be cited as one such minor fruit which is rich in pection of highly jelly-grade and hence, makes better jelly than the common guava fruits.

Chapter 5

Propagation Methods Followed

5.1 Types of Propagation that are Adopted in Fruit Plants

A number of methods of propagation are in fact, adopted in fruit plants. All these methods are broadly classified into two main groups, which are (i) *sexual* and (ii) *asexual* (syn. *vegetative*). In sexual propagation, *fertilized embryos* are used that are present only in *seeds*. In asexual method, plant parts other than fertilized embryos, *i.e.*, *vegetative parts* are used.

Majority of the fruit plants come under *highly cross-pollinated* group. Hence, they are *heterozygous in their chromosomal make up*. Accordingly, the plants that are produced by them, *i.e.*, from the fertilized embryos do not produce plants which are true-to-type to the mother plants in all respects.

In other words, all the morphological, anatomical, cytological or genetic traits (characteristics) are not retained (maintained) in the offspring as are present in the mother plants. This variation that occurs in the offspring are due to *chromosomal heterozygosity* that occurs in them as a consequence of *high degree of cross-pollination*. On the other hand, where high degree of self-pollination occurs in the flowers of a plant, homozygosity results in the chromosomal constituency and as a result, all the characters or most of the characters are *retained in the offspring of the mother plant*. However, it is observed that majority of fruit trees or shrubs are cross-pollinated and accordingly, the plants produced from their fertilized embryos being heterozygous do not retain all characters in the offspring. For this purpose, these fruit trees or shrubs are propagated by asexual means, in other words, by the use of vegetative parts.

Vegetative method of propagation may be of different types and these have been stated below with short description of each method.

1. Rootage

In this method of vegetative propagation, *adventitious roots* are made use of. By adventitious roots is meant production of roots from such a part in the plant body which is not the normal position of root formation, that is, *not from radicle* (root axis). Adventitious roots are produced either naturally in some plant species, or these roots may even be *forced to form artificially* by adopting some treatments and with application of suitable rooting medium on the concerned part.

In consideration to the forced production of roots by adopting treatments and application of rooting medium, two types of propagation methods are employed and these are called (i) cutting (or cuttage) and (ii) layering (or layerage). The difference between these two methods lies in the fact that in case of (i) cutting, plant parts are at first cut off (= detached) from the plant and then, after adopting special treatments to them, they are planted in suitable *rooting medium* when they strike roots. On the other hand, in (ii) layering, at first the adventitious roots are forced in the concerned part of the plant by applying treatments and rooting medium. After formation of adventitious roots in the concerned part, it is then cut off from the mother plant. In horticultural terminology, the rooted cutting is called *cutling* (or cutting) and a rooted layer is called a *layerling* (or layer). The rooted cutting or layer is planted in suitable place to grow and assume the fully form tree or shrub.

Plant parts that are used for cutting may be stem, root, leaf-bud or leaf (= lamina) but stem is more commonly used. Again, according to maturity status of stem, the stem-cuttings are grouped as *hardwood, medium hardwood, softwood* and *greenwood cuttings*. In regard to layering, it is performed on stem part only. In fact, layering may be of two types. These are, (i) when adventitious roots are forced to appear on any *aerial part of the stem* of a tree or a shrub by the application of suitable rooting medium on the concerned part of the stem, it is called *air-layering* (= marcottage) and (ii) when such roots are forced to from using the *ground* as the rooting medium, the method is termed as *ground-layering*. That is, in case of air-layering, it becomes necessary to apply rooting medium on the stem part to form roots while in ground-layering, it is not necessary to apply rooting medium and the soil itself is used as the rooting medium. However, it would be of advantage if the soil is enriched by applying suitable organic manure. There are various methods of ground- layering and some examples are *mound* (syn. stool), *tip, trench, serpentine* layering *etc.*

Time of Preparation of Cutting and Layering

Best time for production of adventitious roots takes place when the meristematic activity of the plant part is high and there is high moisture in the atmosphere. In this sense, rainy-season is best considered. Some nurseries set the cuttings in *mist* chamber so that high relative humidity in the atmosphere is maintained which in having high relative humidity is conducive to root formation.

2. Graftage

In this method of propagation, the shoot of a plant or only one vegetative bud of it is propagated and this is done by artificially *conjoining it with another plant* which is already in rooted condition. The rooted plant on which the shoot or the vegetative bud (which is called *bud-piece*) is joined is called *rootstock* (syn. *stock, understock*) while the shoot or the bud-piece which is made to join over it and encouraged to grow is called *scion* (syn. *cion*).

Grafting may be of two types, which are, (i) *scion-grafting* or, *grafting* and (ii) *bud-grafting* or, *budding*. In scion grafting (= grafting), a shoot which contains more than one vegetative bud, the wood, bark and the terminal bud is joined (= grafted) over the *rootstock plant*. In bud-grafting (= budding), only one vegetative bud which also contains a small piece of bark and a de-bladed petiole is joined (= budded) over the rootstock plant. The vegetative bud is prepared in budding by removing the wood, de-blading it so as to retain a short piece of petiole along with the bud and a short piece of bark and this is called a bud-piece. Due to grafting of budding process, the scion-shoot on the bud-piece respectively unites completely with the rootstock in their combial tissues in such a manner that, the *stion* (= stock, *i.e.*, rootstock+scion) behaves as one plant and freely grows.

Scion-grafting (= grafting) may be of two types. These are, (i) grafting *by approach* (= attached method and (ii) grafting *by separation* (= detached method). The difference between these two method lies in that, in grafting (i) by approach, both the rootstock and the scion grow on their own roots at the time of doing the grafting operation. When union (= union of cambium tissue) between the scion and the rootstock has taken place, the graft is cut off (= separated) from the mother plant (= mother rootstock) and allowed to grow as a single unit, *i.e.*, a single plant. On the other hand, in case of grafting (ii) by *separation* method, the scion shoot is at first cut off from the mother plant and after that, it is grafted over the rootsock plant. Example of attached method of grafting are inarching, saddle-grafting etc. while examples of detached method are whip, tongue, side, veneer, notch, epicotyl (= stone) grafting and besides these, there are also other methods.

It should be noted that in case of grafting, the scion shoot consists of (i) wood, (ii) one or more vegetative buds and (iii) the optical bud. On the other hand, in all budding methods, only one (i) vegetative bud (bud-piece), (ii) a small piece of bark and (iii) a short piece of de-bladed petiole constitute the scion. Wood is never left in the bud-piece except in one method which is known as *chip-budding*. In this method, a small chip of wood is left and make it attached to the bud-piece.

Time of Carrying Out Budding and Grafting

Best time of doing these operations is when the *cambial flow* in the rootstock and scion become high. Rainy-season and spring are the best seasons to perform but in rainy-season, rain water may go in the operated part and does not bring success. Hence, in high rain fall area, these should be done at the end of rainy season and with care. In case of many fruit plants, success of budding is found to be high when done in spring.

3. Micropropagation

In this innovative method, plantlets are made to regenerate from a small part, *i.e.*, tissue which is taken out by cutting from the plant. The cut out part is grown, *i.e.*, *cultured* in a *suitable medium* and under strictly aseptic condition. The medium is fortified by addition with plant nutrient salts, plant growth regulators *etc.* and proper reaction of the medium as well as its moisture content are carefully maintained. At first, *callus tissue* (= simple parenchymatous tissue) develops in the medium from the small (= micro plant tissue kept in the medium and subsequently, new plants (= plantlets) with root and shoot are produced. The plantlets formed are grown, *i.e.*, transferred in suitable media to make increase of the plantlets and then they are planted.

4. Vegetative Embryos

It has been stated that a completely developed seed must contain a *fertilized* embryo. In case of highly cross-pollinated plants, chromosomal heterozygosity occurs and hence, the plant which is produced from a fertilized embryo do not retain all characters as with the mother plant. But in case of some plant species, apart from the fertilized embryo, one or more embryos are also produced vegetatively. The plants that are produced from these vegetative embryos (= apogamic embryos) develop from *nucellus* tissue and these plants maintain all characters of the mother plant. Hence, when propagation is done from these embryos, it could be regarded as asexual, i.e., vegetative propagation method.

In mango, such type of embryos are present in few varieties, *viz.*, Olour, Bappakai, Kurakkan, Carabao, Chandrakaran *etc.*, which are not commercially important. The phenomenon however, common in all *Citrus* species, except shaddock, *i.e.*, pummelo (*Citrus grandis*), citron (*C. medica*) and Tahiti lime (*C. latifolia*). In fruit plants belonging to some other species, nucellar embryos are found to be present.

5.2 Advantages and Disadvantages of Asexual (Vegetative) Propagation

It has been stated under the Heading, 5.1 that majority of the fruit plants are highly cross-pollinated plants and hence, they are highly heterozygous in their chromosomal constitution. Due to this fact, the fertilized embryo which is present in the seed of them produce plant that does not become true-to-type to the mother plant in respect of all characters. For the above reason, it is recommended that fruit plants should be largely propagated by vegetative means and not from seeds. However, it may be mentioned in this connection that although the fertilized embryo is present in the seed, in some species of fruit plants but not in all, non-fertilized, i.e., vegetative or nucellar embryo(s) is also present in the same seed apart from the fertilized embryo. Thus, the nucellar embryo(s) gives rise to true-to-type plant unlike the fertilized embryo. Hence, seed propagation should always, *i.e.*, in each and every species of fruit plant should not be considered as sexual propagation. However, vegetative embryos are present in the seeds of only in few species of fruit plants, Besides that, identification of the fertilized and the non-fertilized embryo(s) becomes difficult to recognize and hence, its (their) dissecting out, *i.e.*, excision for the purpose of propagation becomes difficult. Accordingly, vegetative propagation in fruit plants by the methods of rootage and graftage as stated under the Heading, 5.1 are largely recommended.

Advantages

It has been amply stated in the above discussion that true-to-type plants could be obtained when propagation is done by vegetative methods. Evidently, this is of great consideration in case of fruit crops and no doubt, if a commercial orchard is desired to be set up. Apart from this however, there are many other advantages of vegetative propagation as stated below.

Vegetatively propagated plants tend to become uniform in growth. In case of seedless fruits, *e.g.*, banana, seedless grapes, pineapple *etc.*, it is not possible to propagate without vegetative methods. Fruit plants propagated by vegetative methods produce flowers and fruits at relatively earlier maturity. There are also many other advantages which are of botanical and horticultural importance. However, a great advantage is to secure the benefit of rootstocks. Rootstocks play a profound influence on the scion part of a plant. Hence, in some fruit plants, desirable rootstocks are used. For example, in apple, M-9, M-4, M-7, MM-106, MM-111 and Merton-792 rootstocks are used respectively to produce dwarf, semi-dwarf, semi-vigorous and vigorous apple trees. Similarly, for sour lime, rough lemon appears to be a suitable rootstock, Rangpur lime proves to be a promising rootstock for sweet and mandarin oranges, troyer citrange, karna khatta *etc.* are known to be diserable rootstock for kinow tangor and the like.

Disadvantage

Despite other advantages, there are also disadvantages of vegetative propagation. Vegetatively propagated plants are less strong and robust and are relatively short-lived than those which are sexually propagated and that their productiveness per tree is also lesser. Being much compact, *air-drainage* in the inner part of vegetatively propagated trees is much hampered. For compactness, production of under-sized cull fruits is also more in them. A great difficulty sometimes arises in grafting and budding in respect of *incompatibility of rootstock and scion*. Apparently it seems after grafting and budding that the rootstock and the scion have united but later it may be seen that they have loosened apart. Loosening which results in detachment of rootstock and scion may take place even after few years of growth, which is termed as delayed compatibility.

5.3 Advantages and Disadvantages of Sexual Propagation

It should be remembered that although seed propagation could not be compulsorily considered as sexual method of propagation in all species of fruit plants owing to the presence of non-fertilized embryo(s) but for majority of fruit plants, propagation by seeds is recognized as asexual propagation, in as much as non-fertilized embryo(s) is found to exist in few species of fruit plants only. Hence, seed propagation and asexual propagation are interchangeably used terms in fruit culture.

Advantages

Plants that are produced from seeds become strong, robust, highly vigorous and live longer than vegetatively propagated plants. Their roots are highly spreading, robust and hence, they do not fall off easily by high wind. Inside part of the trees or shrubs get optimum amount of air and light. Branches are large in number, robust and productive capacity of a seed-borne tree or shrub is higher than one which has been vegetatively propagated. In case of some fruit plants, *e.g.*, papaya, phalsa, date-palm, mangoesteen, cape-gooseberry *etc.*, well- built plants for growing as fruit crops are difficult to obtain from vegetative propagation and hence, seed propagation becomes a must.

Disadvantages

Evidently, true-to-type plants do not appear when propagated by seeds in many fruit plants but not in all species. Time to attain bearing maturity, *i.e.*, juvenile or gestation period is more as compared to vegetatively propagated plants. Growth of a tree is not uniform and it occupies large area. In seeds of some species, *e.g.*, citruses, cherry, peach, almond *etc.*, viruses remain present in the seeds and these are transmitted to the resulting trees or shrubs.

5.4 Recommended Propagation Methods for Fruit Crop

There are many fruit plants which are possible to be propagated in a number of ways. But it is often noticed that suitability of all the methods in a fruit plant is not same, besides that, productiveness and other characters are also influenced by the propagation method employed. Hence, for large-scale or commercial fruit culture, some particular method are usually recommended. The recommended methods in case of some fruit crops are stated below. However, it needs pointing out that which method(s) of propagation should be ideal for a particular species of fruit plant in a specific edapho-climatic situation is needed to be confirmed by experiments.

1. Mango

Mango is conventionally propagated all over India by *Inarching* (= approach-grafting) method. The method is cumbersome as thatched roofing becomes necessary to be made for lifting the potted rootock plants above the ground level. The author [(Ref. Mazumdar, B.C. 2022. Improved method of inarching in mango. *Krishi Jagran* (*Engl.*), N. Delhi)] suggested an improved method by which the scion tree could be crept on ground artificially by forcing and thereby, the operation could be done, sitting on the ground, *i.e.*, without using thatched roof.

The fruit scientists in India have recommended some other methods (= detached grafting and budding) to replace inarching. These are, veneer-grafting, chip-budding and Forkert-budding. For all grafting and budding methods, rootstocks have to be prepared from fully mature, plump stones obtained from wildly growing, vigorus mango trees. Grafting should be done at 0.5-0.9 metre height of rootstock. It should be performed in July-August in north India, July-February in south India and end of rainy-season in east India.

2. Banana

Suckers are used for propagation of banana. The suckers along with underground bulbs are cut off from the bulb of the mother plant and these should be large (60-90 cm) in size. Two types of suckers are produced from banana plants, known *as sword suckers* and *water suckers*. Between these two, sword suckers are considered superior. Sword suckers have large base, gradually tapering to the tip and look like sword. Plants that come out from sword suckers grow vigorously and attain bearing maturity early. Water suckers usually arise from diseased rhizome. In Maharashtra (East khandesh area), the Basrai dwarf variety of banana is propagated from dormant rhizomes. Plants raised from micro-propagation method (*in vitro* culture) are also effective materials to obtain uniform and disease free plants of banana.

3. Citruses

A number of propagation methods are followed for the purpose of propagation of different *Citrus* species. These include seed propagation, medium hardwood and hardwood cuttings, layering and budding. Each of these methods has merit in the respective zone. Rootslock has a great influence on scion in citruses. Rootstocks belonging to a number of Citrus species are used for scions of different species. The most prominent rootstock is *Citrus jambhiri* and next is *C. karna*. Other notable rootstocks are Rangpur lime, sweet orange, Cleopatra mandarin *etc.* The *non-Citrus* species, *i.e., Poncirus* trifoliata (= Trifoliate orange) also makes a good rootstock. For commercial culture of *Citrus* species propagation method and rootstock which should be ideal for the respective zone should be judiciously determined before adoption of propagation method.

4. Guava

Guava plants prepared from air-layering make very good materials. Plants made from mound (= stool) layering also serve as highly effective for commercial culture. However, plants prepared from veneer grafting, Forkert budding inverted 'T' budding *etc.* may also be used for commercial cropping. Although inarching is done in guava but other methods seem to have superiority.

5. Papaya

Papaya is universally propagated from seeds. However, the fungal disease, damping-off very commonly appears in the seedlings. To ward off the disease, the nursery beds should be properly drained and treated with Agrosan, Ceresan, Monsoan *etc.* Papaya plants can be experimentally propagated by cutting and grafting but these plants are not suitable for orcharding.

6. Grapes

Hardwood stem cuttings, which are one-season old, 8-10 mm thick (diameter) and prepared from the cut out stem parts of October pruning are suitable materials for adventitious root formation.

In this connection, it is pointed out that, in all plants, a stem-cutting should ideally contain at least two bold buds. The lower and of the cutting should be cut such that, it remains just below a vegetative bud. Similarly, the upper end should be cut just above a vegetative bud. But in grapevine, the length of internode, i.e., the distance between two vegetative buds is considerably long. Hence, the stem-cuttings of grapes become unusually long.

7. Pineapple

In pineapple plant, many parts could be used for propagation. These are, (i) *crown*, which is a cluster of small leaves borne on the head of the fruit, (ii) *aerial suckers*, which arise from the leaf-axils, (iii) *ground suckers* which arise from the base of the plant, *i.e.*, near the ground, (iv) *slips* which are also called *gills* are shoots which appear on the stem just below the fruit and (v) the *stem* on making it into pieces. It is interesting that maturity of the pineapple plant to produce fruit depends on the propagating material which has been used. That is, the topmost part which is the crown takes maximum time and coming in downward direction, slips take lesser time and in this way, the plant takes still less time to fruit when propagated from aerial suckers and still lesser when propagated from ground duckers. Among the different propagating materials, plants are more commonly raised from aerial and ground suckers. However, the Kew group of pineapples produce less suckers and hence, slips are used for propagation of these plants.

8. Litchi

Plants produced from air-layering are highly suitable materials to make orchards of litchi.

9. Ber

The plants belonging to the species, *Ziziphus mauritiana* and other Indian species of ber are best propagated from 'T' budding, inverted 'T' budding and annular or ring budding done before or after rainy-season or in spring. The Chinese species of ber is more commonly propagated by grafting.

10. Jamun

Jamun is more commonly propagated from seeds but the trees raised from seeds show many variations and produce small-sized fruits. To secure true-to-type plants, Forkert and modified Forkert budding which are practiced in south-east

Asian countries could be recommended. Vegetative propagation may also be done by inarching method.

11. Apple

Apple is conventionally propagated by budding or grafting. Among budding methods, 'T'- budding is done during June with current season's growth and among grafting, whip or tongue grafting is usually done and in spring. Rootstocks which are used consist of seedlings of *grab*- apple and also seedlings or clonal rootstocks of pollenizing varieties, e.g., McIntosh, Golden delicious, Rus pipin (Yellow Newton), Northern spy and also Red delicious, which has few seeds in fruits. Rootstocks greatly infhience growth and vigour of scion as has been mentioned under the Heading, 5.2.

12. Pear

As in apple, propagation of pear is also done by 'T' budding in June-July or whip grafting at the end of winter. Pear species of wild types, *e.g., Pyrus pashia, P. khasiana, P. phyrifolia etc.* are used as stocks. Use of Quince 'C' rootstock make dwarf trees. Quince 'A' rootstock makes standard size. Grafting in pear has a difficulty in that, many rootstocks are *not compatible* with the scion. Experiments done with *double-grafting* with some varieties have been found to ensure stionic compatibility.

13. Peach

Peach is propagated by budding or grafting. Seedlings of peach are usually used as rootstocks but seeds need moist *stratification* for easy germination. Sometimes, seedlings of plum, almond or apricot are also used as rootstocks to graft peach scion on them.

14. Plum

Plum is propagated by 'T' budding or also grafting. Seedlings of wild apricot are much used as rootstock. Myrobalan 'B' plum stock (= clonal) is also used. Seedlings of wild species of peach make better rootstocks in light soil.

15. Cherry

Cherry of sweet species (*Pyrus avium*) is propagated by whip or tongue grafting using wild species of cherry known as Paja (*Prunus cerasoides*) as rootstocks in Kashmir and northern Himalayas.

16. Strawberry

The herb is propagated by runners which are one year old. A plant may produce more than 10 runners. Propagation may also be done from crowns.

Chapter 6

Demarcation of Area and Protection against Tresspassing

6.1 Necessity of Limiting Area

After the land has been possessed for establishment of the orchard, next immediate step for the orchardist should be to make necessary arrangement for *clear demarcation* of its area.

The importance of this step lies in the fact that, the actual area of land acquired by the orchardist, in other words, the boundary line of it sometimes becomes an issue of dispute with owners of the land, roads, tanks, *etc.* which are located adjacent to it.

Such type of disputes may arise soon after the land has been acquired by the owner, or even many years after that. In some localities, the danger of gradual encroachment of the land by dishonest owners of the adjoining properties cannot also be ruled out.

It should be remembered that conflicts or ill-feelings with neighbours in respect of area of land is never a desirable happening in orcharding. Therefore, serious attention needs to be paid by the proposed orchardist to limit his land by signifying clearly its boundary line. Beyond doubt, this should be done as quickly as possible after legal possession of the land, because delay in taking action may unnecessarily lead to a number of complications.

In designating the limit of land, the owner should particularly consider two aspects. These could be stated as follows.

(i) It should be located at such position that the owners of the adjoining properties become satisfied, *i.e.*, they have no point to raise any ral or lawful grievance, and

(ii) It should be clear and perspicuous to such an extent that it is easily recognizable not only by the owner himself and the owners of the adjoining properties but by any third person also who has a sense of justice.

6.2 Provision of Protective Measures at the Boundary

In some localities, regular trespassing into the orchard land by human beings, or intrusion of grazing or wild animals into it appear to be a troublesome affair.

These undesirable elements may cause great damage or pilferage of the crops or other assets of the orchard. Besides, conflict with knaves may also result, concerning the unlawful activities by them.

It should not be forgotten that the petty verbal duel of today may turn into a violence tomorrow. Such unwelcome events may badly act as serious impediments in smooth running of the orchard. Hence, adequate steps to protect against this bottle-neck need to be adopted before it is too late, *i.e.*, before any other operation is taken up in the orchard.

As a matter of fact, when safeguard measure along the boundary line of the orchard is properly done, it also demarcates the area by itself. Thus, any separate arrangement to limit it usually does not become necessary.

6.3 Factors Affecting Selection of Protective Measures

There are in fact, a number of devices for protecting the orchard area against unlawful entry of man or animal. These have been enlisted and discussed under the next Heading. It should be kept in mind that each of these methods has its own merits and demerits.

Thus, Heading the method which is actually to be adopted by the orchardist could not be regarded to be the same everywhere and would depend upon a number of factors. These have been stated below.

(i) The expenditure involved evidently differs according to the methods and hence, funding capacity of the orchardist is beyond doubt, an important consideration.

(ii) Whether the requisites that are required to construct a protective barrier are easily available or not should be ensured.

(iii) Attitude of the owners of the properties adjoining the land.

(iv) Infestation of pilferers and anti-social persons in the locality.

(v) Area of size and shape of the land.

(vi) Species and varieties of fruit crops that are to be grown in the orchard.

(vii) Whether inter-cropping with other crops, or mixed-fruit cropping is to be done in the orchard.

(viii) Wether mixed-farming is to be done, such as pisciculture which requires tank, or other forms of animal farming.

(ix) Soil-erosional problem, if any exists in the land.

(x) Threat of storm or cyclone in the locality.

(xi) Personal choice of the orchardist himself.

6.4 Protective Methods: Their Construction, Merits and Demerits

1. Live Hedge

When a row of bushy plants is grown at the boundary of any area for the purpose of its protection against unwanted entry of man or animal into it, or for decoration as well as protection (as the case may be in ornamental gardens, precincts *etc.*), it is referred to as a *hedge* or a hedge-row, or is also termed as a live hedge to distinguish it from other systems, where non-living materials are used.

It should be noted that a row of bushy hedge is able to provide protection of the area to some extent only but not thoroughly. Therefore, to ensure greater protection, it is advisable to provide more than one row of hedge with bushy type of trees or shrubs and these should be planted closely together, such that a more compact and thereby, *impenetrable* form of the thicket could be established. Such an effect is usually not obtained when only a single row of hedge is maintained.

(a) Factors Affecting Selection of Plant Species for Hedge

There are in fact, quite a large number of plant species having the suitability for use as hedge. Those which are of more common use have been enlisted under the next Heading (b). In this connection, it is pointed out that in setting hedge-rows at the boundary, *defence should be given emphasis in a commercial orchard, rather than decoration*. With this consideration, the following factors should be given attention to in selecting what plant species should be preferred for this purpose.

(i) Presence of thorns or spines in the hedge plants is no doubt, a very important criterion. A thorny or a spiny bramble is definitely more effective in resisting intrusion of a man or an animal than one having no prickles. Thorniness is again, superior to the spiny plants.

(ii) Effectiveness of a hedge determines much on its spreading habit. Hedge plants which have highly spreading foliage parts assume more bushy in appearance and therefore, is more impenetrable to the intruders.

(iii) In choosing a species or variety of a plant species for use as hedge, its edapho-climatic suitability under local conditions is no doubt, an important factor. Evidently, a hedge plant which is not well-suited to the soil and climate of the locality where it has been decided to be planted could not be regarded to be an effective material.

(iv) Hardiness of the hedge plants is also a determining factor and the plants which have greater resistance against soil and climatic adversities are more befitting for use.

(v) Rate of growth of the plant species is another important factor in selecting a hedge. A fast growing hedge forms the row in a relatively short time than one, which makes slower growth.

(vi) Wether the row of the hedge has any other economic benefit is also an aspect that deserves attention. There are some hedge plants, *e.g.*, karonda, wild ber, mulberry, cranberry *etc.* which produce fruits having edible value to the human beings, either in fresh form or after processing. Accordingly, an extra income is possible to be obtained by selling the fruits harvested from these hedge-rows.

Some ornamental trees like China rose, *Tabernemontana, Nyctanthes* etc. are also sometimes used to make tall hedges, or low-heighted windbreak rows. Since flowers of these trees are marketable, an extra income is possible to be earned from these hedges also. Again, mulberry trees, if used as hedge have their utilitarian value in rearing silk-worms.

Hence, if any earning from the hedge is desired and even for fuelwood, the species of plants should also be selected accordingly.

(vii) Resistance of the hedge plants against insect or animal pests and diseases should also be duly considered in selecting the species. This is because; hedge-rows of some species harbour insect pests or diseases, which may infect the fruit trees of the orchard. Such type of hedges make great harm to the fruit trees and should not be preferred.

(b) Plant Species Used as Hedge

There are in fact, quite a large number of plant species, which have suitability for use as hedges under particular edapho-climatic and social conditions. Some of the commonly used species have been listed below. It should be noted that the trees or shrubs belonging to many of these listed species become much taller and may even serve as low-heighted windbreak rows. However, to maintain them as hedges, regular pruning of them is necessary.

Hedges having Throns or Spines

(i) *Acacia arabica* and other *Acacia* species, *viz., A. concinnao, A. farnesiana, A. latronum, A. longifolia, A. tortuosa etc.* (Eng. and Ind. names: Gum tree, *Babul, Babla, Baburi, Babhul, Kikar, Karuvelam, Nalatuma etc.*). These are very much suitable in dry region.

(ii) *Agave americana* and other *Agave species.* (Eng. and Ind. names: Century plant, American aloe, *Kantala, Jangli anaras, Kolakantalo,Parkanda, Ghaypat, Nattukaita, Bontharakasi etc.*). Highly protective against instrusion of animals.

(iii) *Bauhinia acuminata (*Eng. & Ind. names: Mountain ebony, *Kanchan, Kachnar* etc.). Have decorative value also.

(iv) *Berberis darwinii*

(v) *Bougainvillea spectabilis* and *B. glabra* (Eng. & Ind. names: *Bougainvillea, Bagan bilas, Kagaj phool* etc.). Shrubs are particularly suitable for training over wire-trellises. Have decorative value also..

(vi) *Carissa carandas* (Ind. names: *Karonda, Karekayi, Karamcha, Karavanda, Kalakkay, Vaka* etc.). Fruits may be used for drying or making other food products.

(vii) *Duranta plumieri (*Eng. & Ind. names: *Duranta, Nilkanta, Nilambari, Ranaboli, Nakkatoka* etc.). Have decorative value also. White or purple flowers and variegated varieties are especially attractive.

(viii) *Euphorbia antiquorum* and other Euphorbia species (Eng. & Ind. names: *Cactus, Tesiramonsa, Tidhara sehud, Dokana siju, Chaudhri niwdung, Chathirakkali, Bonthakali, Shadraikali* etc.). Succulent shrub.

(ix) *Inga dulcis* (Eng. and Ind. names: Manila tamarind, Madras thorn, *Jilebi, Vilayti imli etc.*). Have decorative value and very suitable in dry region.

(x) Lantana camara (Ind. names: *Chotra, Ariple, Nagairi, Ghanidalia, Ganeri, Puchedi, Arippu* etc.). Very suitable in dry region.

(xi) Lawsonia alba (Eng. & Ind. names: *Henna*, Mehendi etc.). Leaves are used for cosmetic purpose.

(xi) *Lawsonia alba* (Eng. & Ind. names: *Henna, Mehendi* etc.). Leaves are used for cosmetic purpose.

(xii) *Opuntia dilleni* (Eng. & Ind. names: Prickly pear, *Phanimansa, Nagphani, Nag-neval, Phadya niwdung, Elakkali, Sappath thikalli* etc.). Very suitable in dry region.

(xiii) *Parkinsonia aculeata* (Eng. & Ind. names: Jerusalem thorn, *Bilati babla, Kikar* etc.). Fast growing tropical plants, suitable in dry region.

(xiv) *Prosopis juliflora* (Eng. & Ind. names: Honey locust, Mesquite, *Vilayti kikar* etc.). Used as fire-wood also.

(xv) *Rosa multiflora (*Eng. & Ind. names; Wild rose, *Jangli gulab, Kath gulab,* etc.). Suitable for training over wire-trellises. Have decorative value also.

(xvi) *Ziziphus mauritiana* and other *Ziziphus* species (Eng. & Ind. names: Indian plum, *Ber, Kul, Ban kul, Barkoli, Boradi, Elanthai, Kulvali, Regu* etc.). Fruits are consumed as raw or products are made. Trees are used in culture of lac insects. Very suitable in dry region.

Hedges of other Species

(i) *Acalypha indica.* (Also have decorative value).

(ii) *Adhatoda vasica.* (Leaves have medicinal value).

(iii) *Annona squamosa.* (Fruits are edible. Leaves are not eaten by animals. Shrub is highly resistant against insect pests and diseases. Suitable in very inferior soil).

(iv) *Aralia species.* (Also have decorative value).

(v) *Areca lutescens.* (Also have decorative value).

(vi) *Bambusa nana.*

(vii) *Barleria cristata.*

(viii) *Berberis darwinii.*

(ix) *Buxus sempervirens.*

(x) *Casuarina equisetifolia. (Also have decorative value).*

(xi) *Choisya ternata.*

(xii) *Codiaeum variegatum.*

(xiii) *Cotoneaster simonsi.*

(xiv) *Clerodendron inermi.* (Also have decorative value. Odour of leaves repel animals).

(xv) *Cryptomeria japonica.*

(xvi) *Dodonaea viscosa* (Also have decorative value).

(xvii) *Daedalacanthus nervosus.* (Also have decorative value).

(xviii) *Eranthemum species.* (Also have decorative value).

(xix) *Euonymus japonicus.*

(xx) *Galphimia gracilis.* (Also have decorative value).

(xxi) *Grewia asiatica.* (Fruits are edible. Tree is very hardy).

(xxii) Grevillea robusta. (Also have decorative value).

(xxiii) *Hamelia patens.* (Also have decorative value).

(xxiv) *Hibiscus rosa-sinensis.* (Also have decorative value).

(xxv) *Ipomea carnea.* (Leaves are not eaten by animals. Very hardy plant).

(xxvi) *Ixora species.* (Also have decorative value).

(xxvii) *Justicia gendarussa.* (Also have decorative value.)

(xxix) *Ligustrum ovalifolium.*

(xxx) *Malpighia coccigera* and *M. glabra.*

(xxxi) *Murraya exotica.* (Also have decorative value),

(xxxii) *Nyctanthes arbor-tristis.* (Also have decorative value).

(xxxiii) *Pittosporum tenuifolium.*

(xxxiv) *Poinsettia pulcherrima.* (Also have decorative value).

(xxxv) *Polyalthia longifolia.*

(xxxvi) *Pongamia glabra.*

(xxxvii) *Plumbago capensis* (Also have decorative value).

(xxxviii) *Prunus laurocerasus.*

(xxxix) *Prunus padus.* (Fruits are edible).

(xl) *Prunus rufa.* (Fruits are edible)

(xli) *Prunus granatum.* (Fruits are edible).

(xlii) *Putranjiva roxburghii.* (Also have decorative value).

(xliii) *Rhapis excelsa.* (Alsoave decorative value).

(xliv) *Rosamarinus officinalis.*

(xlv) *Rubus* species. (Fruits are edible).

(xlvi) *Sesbania aegyptiaca.*

(xlvii) *Tabernaemontana coronaria.* (Also have decorative value).

(xlviii) *Tamarix pentandra.*

(xlix) *Tecoma stans.* (Also have decorative value).

(l) *Thevetia nerifolia.*

(li) *Thuja species.* (Also have decorative value).

(lii) *Thunbergia erecta.* (Also have decorative value).

(c) Establishment and Maintenance

In an orchard, an ideal hedge plant should be such a one which is robust, hardy and has highly spreading branches starting from near the ground level. That this such that when grown up, it would have a compact, bushy and impenetrable appearance. Presence of thorns or spines in it is no doubt, an added advantage. It should be remembered that for building up a row with such type of hedge plants, it is necessary to adopt suitable measures from the very beginning of setting it.

The hedge plants can be raised from seeds or by vegetative propagation methods, depending on the species and choice of the builder. However, the plants raised from seeds are usually found to be hardier, spreading and of more vigorous growth than those which are vegetatively propagated. Again, sowing of the seeds *in situ* gives better result in most cases than transplanting of the seedlings. This is because, in the former case, the transplanting shock to the seedlings does not arise, besides the fact that the roots also ramify and get established in the soil freely in this method.

For seeds of hedge plants of some species, germination appears to be a difficult process but suitable treatments might ease their germination. The treatments have been stated later in the Chapter 9.

Instead of seeds or seedlings, if cuttings are used as propagating materials, these should be rooted in the nursery at first and then the cuttings (rooted cuttings) be planted. This may not be necessary for such cuttings which form roots easily and these cuttings may be planted directly. There are however, cuttings belonged to some species which do not produce roots easily. For these difficult-to-root cuttings, special treatments may become necessary and these have been mentioned in the Chapter 9.

After the seedlings or the rooted cuttings have well-established in the soil,next attention should be paid to ensure that the plants make vigorous and thick-stemmed growth during the first few years. This is a very important criterion for a hedge plant as lanky, less spreading and thin-stemmed plants if are produced in the early part of growth cannot make an effective barrier at later time. To force desirable growth of the plants, the following procedure is however, of much benefit.

A *trench* of about 15 cm depth should be dug at the position of sowing seeds or planting of cuttings. The width of the trench would vary depending on the number of rows of hedge plants that are to be maintained. The trench should preferably be dug in summer season and the dug out soil should be heaped around the trench for hot-weather exposure until the beginning of rainy-season.

After getting weathered, the soil should be thoroughly mixed with adequate amount of farmyard manure and some quantity of nitrogen, phosphorous and potash fertilizers and then the trench should be filled with this soil mixture. Before returning to the pit, it is also advisable to apply some quantity of Aldrin or other insecticide to the soil as a safeguard measure against white ants. The filled up trench should then be watered heavily for few days. If any depression is noticed after watering, it should be made up to the ground level by applying soil manure mixture. Then sowing or planting should be done in it.

Evidently, irrigation should be done regularly after sowing or planting, if there is no rain and the irrigation should be done more frequently but with less quantity. Where seed sowing is done, careful thinning of the emerged out seedlings also becomes necessary so as to maintain suitable spacing.

In order to impart a bushy and compact form of the hedge plants, they should be forced to produce side branches much as possible. This may be done by de-topping and repeated clipping of their branches at early stage of growth. For this purpose, the plants should be allowed to grow unhampered upto first six months or so and thereafter, de-topped and the branches turncated. After this clipping operation, little amount of nitrogenous fertilizer should be mixed with the soil and then watered.

Such type of trimming should subsequently be done once or preferably twice in a year, before rainy-season and then in winter. Application of organic manure, phosphate and part of potash fertilizers should be applied after trimming in winter, while nitrogen along with part of potash fertilizers after trimming in monsoon. The manures and fertilizers need to be mixed thoroughly in the soil and then irrigated heavily. Foliar spraying the hedge plants occasionally with the 'Crop Growth Stimulant' causes to boost up heavy vegetative growth of them in a short time.

Within 2-3 years, the hedge plants will make good growth and after that, manures and fertilizers may be applied only occasionally and when there is necessity.

When the hedge plants have grown sufficiently tall, they should be lopped at the head at desired height. Any growth of shoot above this height should be headed-back, i.e., detopped.

Very often it is notice that some gaps are left at the basal part of the hedge-rows, which is due to lesser growth of branches at this part. Such type of sparse growth does not result, if the plants are properly clipped and well-manured. However, if such gaps are noticed, these should be filled by sowing of seeds or planting of cuttings as the case may be.

Some species of hedge plants create a difficulty in a way that, the seeds borne by them in mature condition fall over the ground around the hedge-row and the seeds being easily germinable produce seedlings abundantly. The hedge plant, *Lantana camara* is a notable example of this type. It is seen that when this plant is used as a hedge, seedlings emerged from seeds of the plant are found even away from the hedge-row. The seedlings go on covering more and more area, thus serving as weeds in the orchard. Regular clipping of the fruit bunches in green stage thus, becomes necessary to avoid the trouble.

For any row of hedge plant, more dense, compact and bushy it assumes, more effective it is in resisting crossing it over by man or animal. Methodical clipping done regularly and judicious application of manures and fertilizers should be the steps to build up such a purposeful hedge. Application of plant protection chemicals as preventing measures is also a necessity in hedge technology to preclude attack of the hedge plants by pests and diseases.

In particular, white ants sometimes appear to be a meanacing problem with hedge plants of some species and the plant species, *Inga dulcis* could be cited to be a good example to that. Hence, soil treatment before setting and application of plant protection chemical in soil and the plants is advocated regularly as preventive measure against invasion of white ants.

(d) Advantages and Disadvantages

Beyond doubt, multiple rows of hedges set closely together affords suitable protection against intrusion of unwanted persons or animals into the orchard to a great extent, thought not completely. Initial cost to build up the rows of hedges is also relatively less than many other methods. Hedges of some species may provide some extra income to the orchrdist as well.

However, it is a point that although initial cost is less, some recurring expenditure is involved to maintain the hedge-rows in proper condition. Another difficulty lies in the fact that some species of hedges harbour insect pests or diseases, which are liable to infect the orchard trees and thus, cause harm to them.

2. Wall

Enclosing an orchard with a wall is sometimes a practice done in many parts of India. A wall may be made up of many materials and the commonly used have been methodsdiscussed in the following.

(a) Brick

A brick wall (Figure 6.1) is built with layers of well-burnt bricks made up of clay, which are suitably joined together with cement, mortar or clay for its support. For anchorage of the wall firmly above the ground, underground foundation work is necessary to be made. Outside plastering of the brick wall is preferred by many, while it is dispensed with by others. The height of such a wall may vary according to the choice of the owner but a height of at least 1.5 metres appears to be necessary to rebuff climbing, clearing or jumping over it easily by a person.

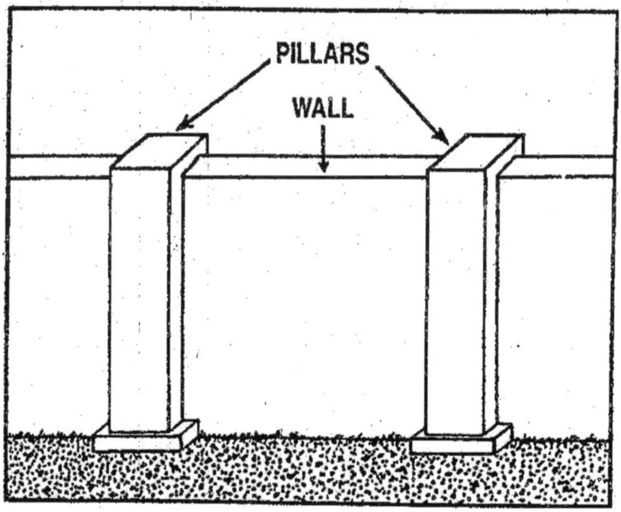

Figure 6.1: Wall made up of Bricks with Outside Plaster.

Instead of making the entire breick wall a continuous one, it is better if it is divided into parts, leaving a small gap of about 2 cm between two adjacent parts. The method has its advantage in that, one or two parts though may be damaged by heavy storm but not the entire wall.

It should however, be noted that a trivially erected brick wall is of little practical value. As a matter of fact, strength, durability and effectiveness of a brick wall depends on a number of factors. These are (i) height, (ii) width, (iii) type of bricks used, (iv) material used in joining bricks, (v) underground foundation, (vi) columns, (vii) outside plastering *etc.*

As shown in Figure 6.2, the effectiveness of a brick wall would be greater if the top of it is made angulated and is suitably embedded with iron nails. By this, climbing or walking over it by a person or animal would be difficult.

In olden days, there had been a practice among many wealthier people in India to surround their orchards or estates consisting of crop farm, orchard, dairy, poultry, flower garden, piscicultural tank, out-house *etc.* with tall and robust brick walls. Nowadays, it has almost become obsolete, perhaps due to very high cost involved in its construction and also maintenance. Nevertheless, for its firmness, impermeability and permanent nature, the method is still followed usually by organizations, institutions, firms *etc.*, rather than individuals.

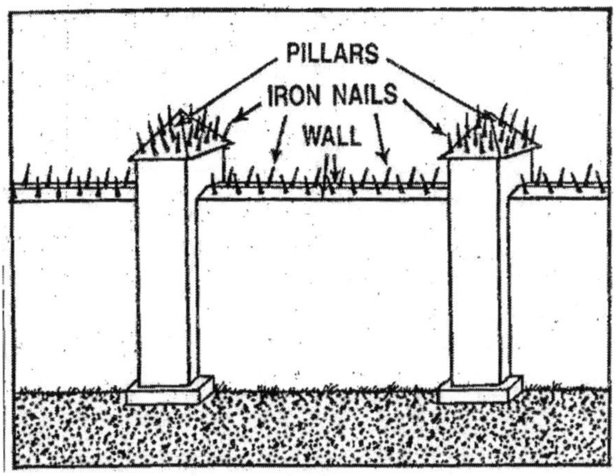

Figure 6.2: Brick Wall with Top Angulated and Embedded with Iron Nails.

Advantages and Disadvantages

A brick wall has evidently, a number of benefits, important of which are as follows.

(i) It makes a clear demarcation of the orchard from its surroundings and thus, rules out contention with others in respect of area of the land that belongs to the owner.

(ii) A strongly built brick wall affords sturdy resistance against human beings, grazing animals and also many wild animals, to whom the wall is not passable.

(iii) Wash-out of surface soil outside the orchard land due to slope, if any at the border area could be fully averted by a brick wall. Thus, a brick wall is a highly effective device to save the orchard soil against erosional loss.

However, a number of demerits as stated below are also there in enclosing an orchard with brick wall.

(i) The cost involved in its construction is very high and is difficult to be afforded by an ordinary orchardist.

(ii) Unless strongly built with high quality and well-burned bricks and concrete mixture, a brick wall may be easily damaged or broken, either naturally or is deliberately done by harmful persons for trespassing.

(iii) High wind is liable to cause damage or felling of a brick wall.

(iv) Another difficulty of a brick wall lies in the fact that the land near the periphery of the orchard remains shaded by the wall and the shaded area may be greater with greater the height of the wall. As a result of shade, the orchard trees which are located near the wall are hampered in their proper growth and productiveness.

(b) Reinforced Slabs

Instead of bricks, a wall may also be constructed with slabs and columns made up of reinforced concrete (Figure 6.3).

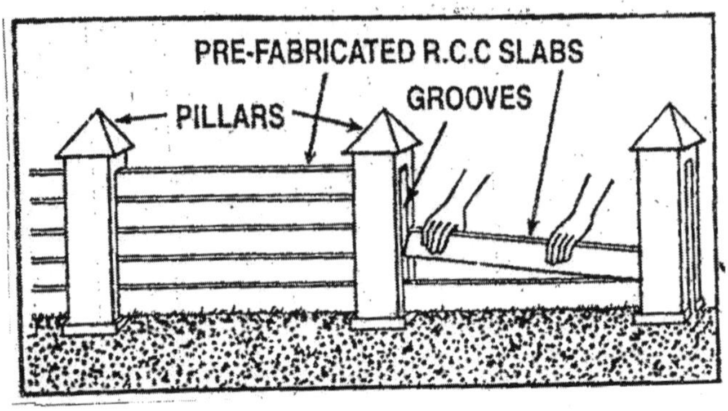

Figure 6.3: Wall of Concrete Slabs.

To construct such a wall, columns with underground foundation works to support them are built, such that desired height of each column remains perpendicularly erected above the ground level. During construction of the columns, grooves are provided on two opposite sides of each column. The columns are erected at a spacing of 1.5-2 metres and in such a manner that the groove of a column lies *vis-a-vis* with that of the next column.

Pre-fabricated slabs or reinforced concrete slabs are made and each slab is vertically placed between two columns. These are placed in such a manner that one edge of it fits in the groove of one column, while the other edge in the groove of the side column. After placing the slab in this way, the next slab is similarly placed over it and the two slabs are then joined together with cement. The slabs should also be cement jointed with the vertical columns. In the same way, the slabs are placed one after the other and joined till the desired height of the columns is attained. As in case of a brick wall, a concrete slab wall should also be at least 1.5 metres high to become effective.

Regarding advantages, it could be said that such a wall is stronger and more durable than a brick wall. Building up wall can be done rather easily and quickly than a brick wall as ready materials are available.

(c) Concrete Lattices

A wall may also be built up with lattices or *zafri*, made up of concrete material (Figure 6.4).

As in case of a brick or a concrete slab wall, in wall of this type also, columns are needed to be erected. The concrete lattices are placed one after the other and all attachments are joined to make a continuous wall as is done in a concrete slab wall.

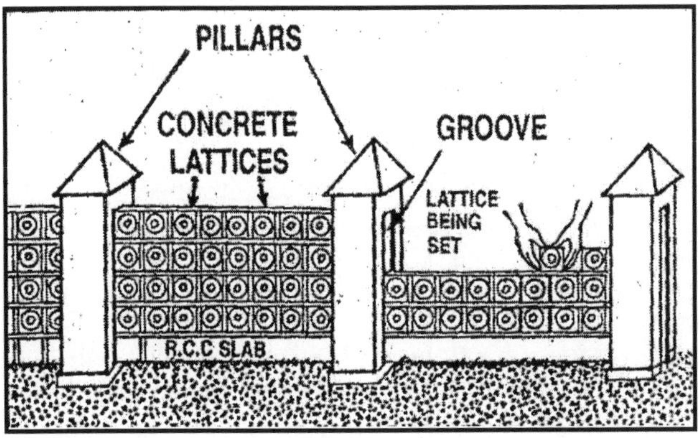

Figure 6.4: Wall made up of Concrete Lattices.

Apart from the protective aspect, a wall of lattices has a decorative value as well. The construction can also be done easily and quickly as ready-made materials are available in the market.

(d) Asbestos or Galvanized Iron Sheets

In construction of a wall, sheets of asbestos or galvanized iron may also be used instead of bricks or concrete. The sheets used may be smooth or corrugated but the latter gives better effect (Figure 6.5).

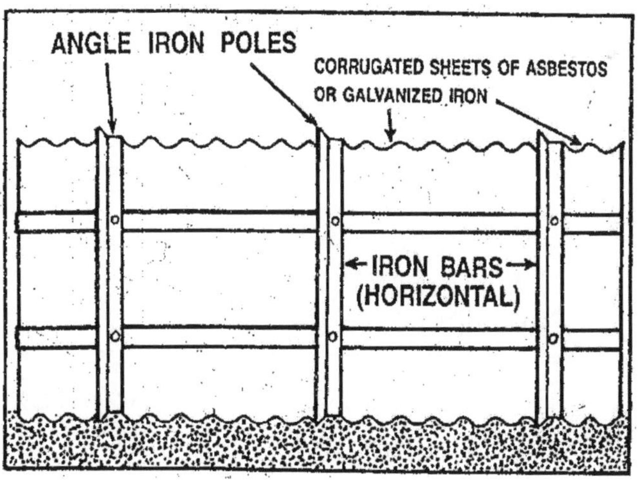

Figure 6.5: Wall of Asbestos Sheets.

To build up such a wall, angle iron bars should be vertically erected, embedding their lower part in concrete platform. Instead of embedding in concrete platform, the bars may also be in fixed deeply into the ground firmly with peeble-stones or by other suitable means. A spacing of 1.5-2 metres should be maintained between two such adjacent vertical bars, *i.e.*, poles. The upper and lower parts of the poles are also joined with horizontally placed bars (It should be noted that as the asbestos is now known to be a carcinogenic material, it is not used by many).

A wall made up of g.i. or asbestos sheets has the advantage in that, it is difficult to be crossed over by an intruder, if it is strongly built. In this respect, it appears to be more effective than other type of walls mentioned above. This is because, climbing or clambering over this wall by a person would perhaps be more difficult than walls of other types.

The wall of g.i. sheets has the drawback in that, the sheets and the angle iron pillars or bars are rusted with time, necessitating their repair and replacement. This can however, be partly averted by coating the materials with coal-tar or suitable paint. If asbestos sheets are used, they are also liable to be broken due to weathering. Enclosing an orchard with asbestos or g.i. sheets has another problem in that, the sheets become very hot in summer and very cold in winter. As a result, the trees in the orchard which are located close to the wall get high heat and cold and thereby, their growth and productiveness may be much hampered.

(e) Stone Pillars

Walls of stone slabs (Figure 6.6) are also found around plantations, at least in localities where these materials are available abundantly with little or no cost as in Karnataka and some other parts of India. To build up a wall of these materials, roughly rectangular blocks of stone, each measuring about 1.5 metre long and 30-45 cm wide are used and these are placed side by side, burying at least 30 cm of each under the ground.

Figure 6.6: Wall of Stone Slabs.

A wall made by stone pillars is no doubt, very sturdy and affords adequate protection against entry of animals. But for intrusion of human beings, such a wall does not appear to be very effective especially because walls of this type where erected are usually made of low height and hence, are easily climbed over by a person. Above all, a wall of stone pillars is difficult to be built up everywhere and could be possible only where these materials are locally available in plenty and at cheap rate.

(f) Mud

Low-heighted mud wall (Figure 6.7) around a farm or an orchard is noticed in almost all over India. To make such a wall, the earth is dug with spade from both sides and heaped so as to form a high ridge. Care should be taken that the earth is not dug from the adjacent land, if the same belongs to other person. The height of a mud wall varies but usually it is not made more than 1.5 metres with a width of 45 cm as a tall mud wall is liable to break more easily.

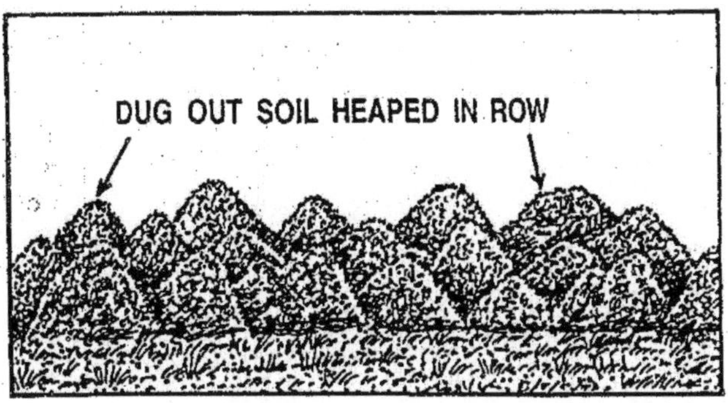

Figure 6.7: Wall of Earth Mound.

A mud wall made along the orchard boundary is satisfactory to afford suitable protection against entry of grazing animals. Above all, it is least expensive as well, compared to other type of walls. Such a wall is however, not strong and may collapse easily with blow or thrust. Rain water also causes washing out of the wall, for which frequent repairing of the wall becomes necessary. Besides, construction of a mud wall is hardly possible with soil, if it is of loose texture.

3. Iron Railing

Surrounding the orchard with railing of vertically fixed iron bars (Figure 6.8) is another satisfactory method to protect it against trespassers. Rounded and pointed iron bars, which are 5-7 cm wide and pointed at one or both ends are usually used. The length of the bars it should be at least 1.5 m to have better effect.

To construct the railing, about 30 cm part of the bars should be vertically buried under the ground. A gap of about 5 cm should be given between adjacent bars. The bars should also be tightly held together by framing with horizontal bars of iron near the lower and the upper parts. Erecting angle iron posts at suitable distances is also desirable to provide support.

An iron railing is sufficiently strong if properly built. It occupies less space of the orchard land and the shade of the border trees by such a railing is also very less. Thrust by wind usually does not fell a railing. The cost is no doubt higher, although lesser than permanent walls.

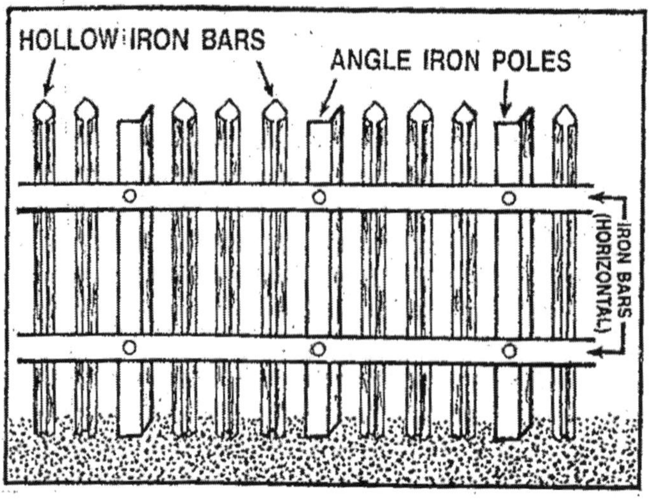

Figure 6.8: Railing of Iron Bars.

4. Wire Fencing

A commonly used system in many parts of India is to enclose a farm or an orchard area with a fence made up of iron wire trellises. The method does not appear to be very satisfactory if used alone but when combined with rows of hedge plants (described later), a protective barrier is formed and at a relatively low cost.

In constructing a wire fence, poles are to be erected at suitable distances of 1.5-2 metres as in case of other methods. The poles may be of angle iron, or reinforced concrete. Wooden posts may also be used when built temporarily. If angle iron poles are used, it is advisable to post them embedding the lower parts of them in concrete platform so that they do not tilt at any side.

The height of a wire fence may vary according to the choice of the owner but it should be at least 1.5 metres for better effect. Accordingly, the poles to be used should be of such length that after embedding in platform or infixing into the ground, at least 1.5 metre length of each pole remains vertically projected above the ground level.

Types of Wire Fence

Wire fencings that are commonly installed are of the following four types.

(i) Parallel, unbarbed (simple) wire fence (Figure 6.9).

(ii) Parallel, barbed wire fence (Figure 6.10).

(iii) Net type (interwoven), unbarbed wire fence (Figure 6.11).

(iv) Net type (interwoven), barbed wire fence (Figure 6.12).

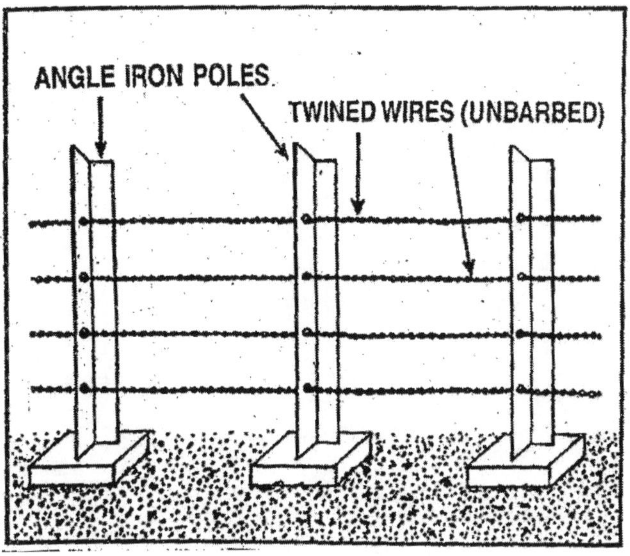

Figure 6.9: Parallel, Unbarbed, Wire Fence.

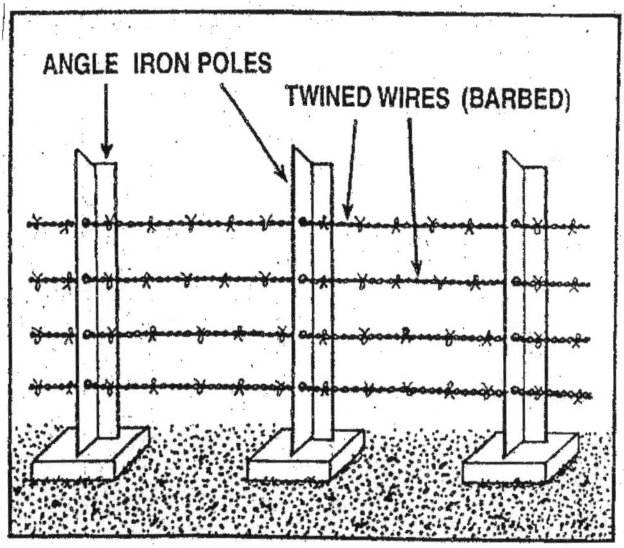

Figure 6.10: Parallel, Barbed Wire Fence.

The wires that are commonly used for these fences are of 0.3-0.5 cm thickness. The effectiveness is no doubt, greater with greater the thickness of the wire. The wires used are of galvanized iron but use of aluminium wire, although costly is better to prevent rusting.

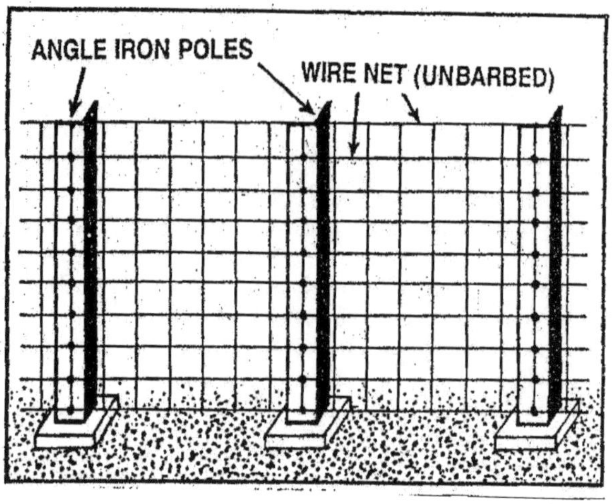

Figure 6.11: Net Type, Unbarbed Wire Fence.

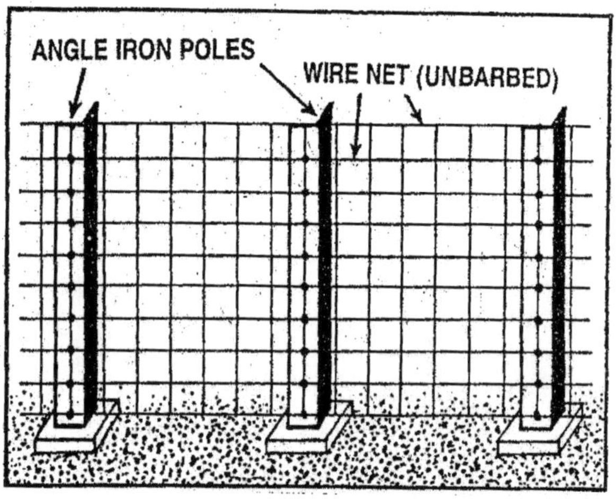

Figure 6.12: Net Type, Barbed Wire Fence.

In parallel wire fence, whether barbed or not as in (i) and (ii), the wires are placed horizontally and held tightly, leaving a gap of 15-25 cm between two wires. The effect would be greater with lesser the space left between them. The wires are firmly tied to the poles with strong binding wire such that they could not be opened apart easily. In this way, the parallel wires should be run upto the height of the posts. In case of a barbed wire fence as in (ii), each wire is provided with barbed spikes at 10-15 cm apart. However, closer gap between the spikes would no doubt, give better protection.

For simple net type of wire fence (iii), nets of wire are used instead of ordinary running wires. The nets may be square as in Figure 6.11, or rhomboid and the side of each square or rhombus may be 8-12 cm. For a barbed net wire fence, the wires that make up the nets are provided with barbed spikes (Figure 6.12).

Parallely run, g.i. wire fencing is of much advantage in affording protection to the entry of grazing animals. Beyond doubt, it becomes more efficient, if barbed wire is used. However, for human trespassers, the method cannot be regarded to be very satisfactory, because one can easily climb over such a fence, using the wires like the rungs of a ladder. To prevent entry of grazing animals in the orchard, the net type of wire fence is much better than the parallel wire type. Again, barbed net gives much better effect than the unbarbed net.

Among other disadvantages are, attempt of the intruders to widen the gap of parallel wires by pulling apart the wires or cutting them with pliers. Rusting of wires with time, thus necessitating their replacement is another recurring expenditure. Wooden posts if used, may also be damaged by the attack of white ants but this problem can be lessened by pre-painting the posts with coal-tar.

5. Dead Wood Fencing

Fencing with very little cost may be done by the use of bamboo slices, with or without combination of clippings of bushes and throns, Such type of fences are however, not of permanent nature and occasional replacement of some parts becomes necessary.

(a) Bamboo Slices

In this type of fence wattle (Figure 6.13), only slices and poles of bamboo are used.

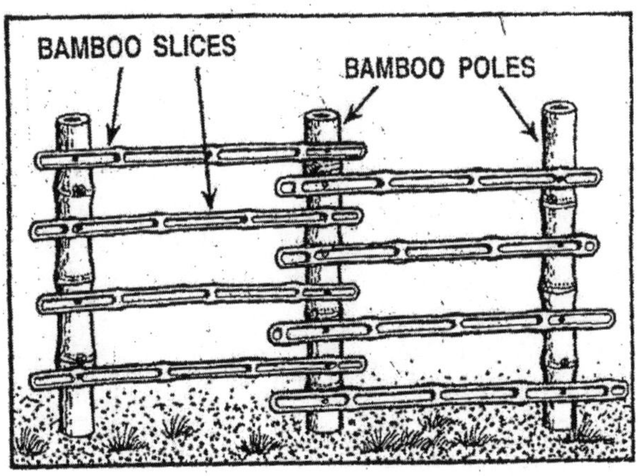

Figure 6.13: Fence of Bamboo Slices.

To construct such a fence, bamboo posts are vertically infixed in the ground firmly to about 30 cm depth and maintaining a spacing of 1.5-3 metres between two adjacent posts. Long slices are made by splitting bamboos lengthwise and the slices, thus made are horizontally affixed to the posts. This may be done by nailing or fastening tightly with wires, one over the other, leaving a gap of 15-20 cm between two slices. The height of a bamboo fence is usually not made high but to be more effective, a height of at least 1.5 metres becomes necessary.

To prevent entry of grazing animals into the orchard, provision of bamboo railing at its boundary appears to be quite satisfactory. But for human beings, it cannot be regarded so, because a trespasser can freely cross the fence by braking it or climbing over it like a ladder, using the horizontally laid slices as the rungs. As stated earlier, a bamboo fence would not also last long as it is easily infected by microbes or get rotten. The durability may however, be increased by coating with coal-tar before furnishing and also at later time.

(b) Clippings of Bushes and Thorns

Fence of this type is similar to a bamboo fence, except that pruned branches of trees, clippings of brunches, thorny trees *etc.* are used instead of bamboo slices (Figure 6.14).

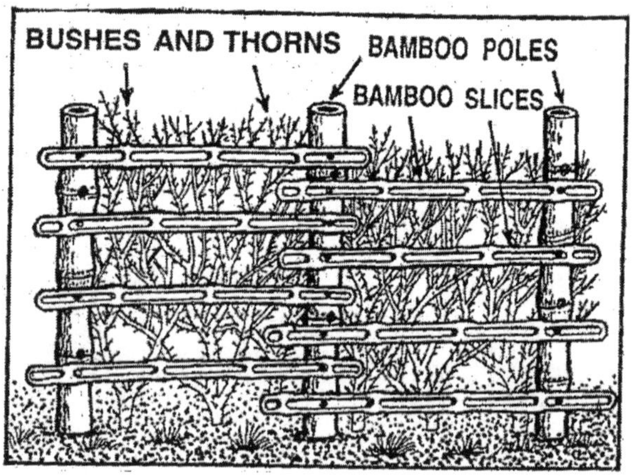

Figure 6.14: Fence made with Clippings of Bushes and Thorns.

The bamboo posts are infixed into the ground as in a bamboo fence. After doing so, the pruned out parts of trees, shrubs *etc.* are infixed to the ground firmly and very closely, leaving little gap between them. Over these pruned materials, clippings are also vertically laid in a thick manner. Finally, a bamboo slice is laid horizontally near the top of the fence and this is nailed or tightly tied to the posts. The bamboo slice may also be tied to some of the clippings. Provision of another bamboo slice near the base is still better such that the clippings are not displaced from their positions.

The frame is cheaper than a bamboo fence and is very effective, especially if thorny branches are used. Due to low cost and effectiveness, this fence is used by the less wealthier people, particularly for enclosing small area. However, it is of temporary nature and needs frequent repairing and restructuring.

6. Trench

Many orchardists in India prefer to dig trenches at the orchard boundary instead of making provision of any other device as the protective measure (Figure 6.15). *The system commonly goes by the term as cattle-proof trench* as in most cases, it is intended to prevent entry of the grazing cattle inside the orchard.

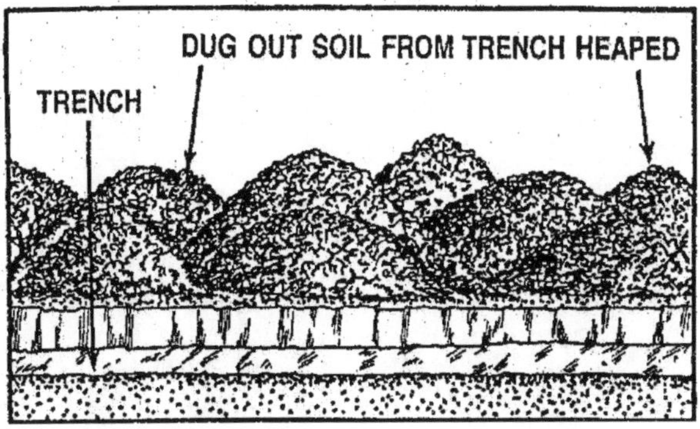

Figure 6.15: Cattle-Proof Trench.

To become effective, the trench should be dug to a depth of about 50 cm and making a width of least 70 cm. The dug up soil should preferably be heaped along the trench to be more effective.

A properly dug trench offers adequate protection against entry of not only cattle but also many wild animals inside the orchard, although it is useless for human trespassers. Occasional repair of the trench by digging so as to maintain the depth and width of the trench may become necessary as it is liable to get filled gradually with the soil, heaped along it due to rain.

7. Ditch

A ditch around an orchard is rarely seen nowadays. But in olden days, many wealtheir people in India had the fancy to dig wide ditches around their estates, which were known by the names as *lahar, nahar, parikha, khat, bandh, dighi etc.* However, their estates usually comprised of fairly large areas and those contained not only orchard but also other features like pleasure-ground, decorative garden, park, play ground, swimming pool, animal farm, angling tank, zoo, pleasure house *etc.*, all at one place. The ditch at the boundary of the estates was dug by them for both security as well as boating purpose and this measured about 1-1.5 metres deep and 3-5 metres wide.

At the present time, making such a moat around an orchard appears to be a ridiculous approach for the very high cost involved in it, besides loss of much valuable land of the orchard. Nevertheless, it seeks to gain ground under certain conditions. For example, for the purpose of irrigation, pisciculture, rearing of water fowls, culture of aquatic crops like water-chestnut, makhna, lotus *etc.* or enclosing estate comprising farms for field crops, vegetables, orchard, animal farm, piscicultural tank *etc.*, a ditch at the boundary (Figure 6.16) may be of some justification, provided the cost of construction is affordable to the owner. If there is any river or canal nearby, the ditch may even be connected to it with lockgate, providing inlet and outlet to the river or the canal.

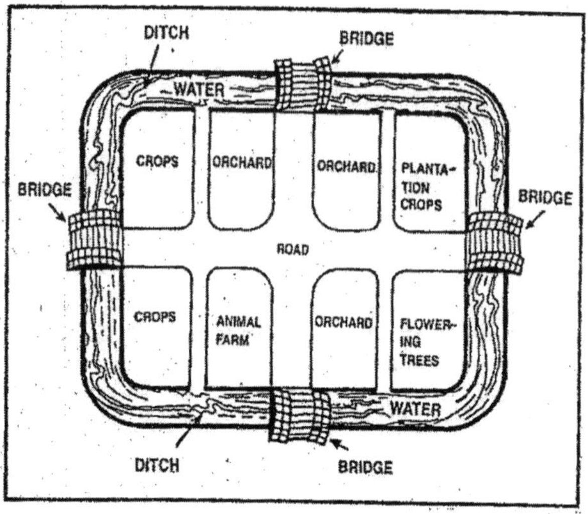

Figure 6.16: Ditch around a Composite of Farm, Orchard *etc.*

Beyond doubt, a ditch provides adequate security of an area against instruction of human trespassers and not only grazing but many wild animals as well.

8. Combination of Two or more Methods

Many orchardists prefer to construct a protective barrier at the orchard boundary by making suitable combination of two or more methods among those which have been stated above. More commonly, two methods are combined together in this system, although sometimes more than two methods are also combined.

(a) Hedges and Wire Fence

A very common as well as effective device is to make a combination of one or more rows of live hedge with that of one running fence of wire. The hedges and the fencings may be among those that have been enlisted under the Heading, 6.4 (1) and (4) respectively.

It is very important that the row of hedge and the running wire fence should be set in such a manner that they are in close touch with each other. The fence may be provided either at the inner side, *i.e.*, towards the boundary and the hedge-row closely attached to this in the outer side (Figure 6.17), or, *vice versa* (Figure 6.18). However, a better system may be to provide one or even more rows of hedge closely in touch at both sides of the wire fence rather than at one side (Figure 6.19).

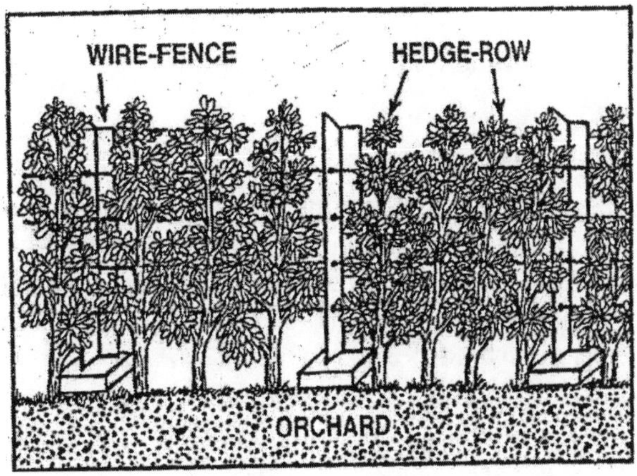

Figure 6.17: Wire Fence Set at Inner Side and Hedge-Row at Outer Side of Orchard.

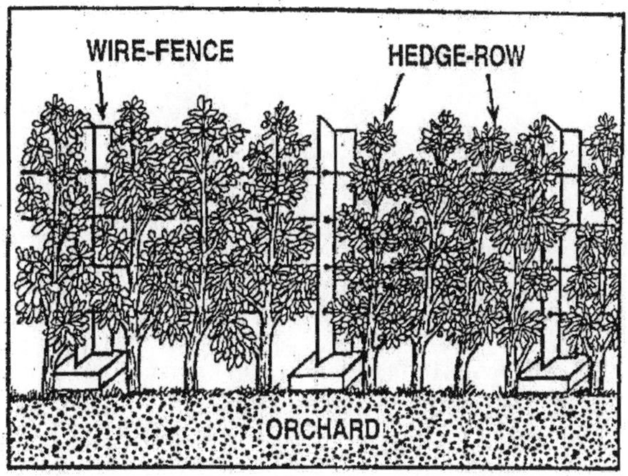

Figure 6.18: Wire Fence Set at Outer Side and Hedge-Row at Inner Side of Orchard.

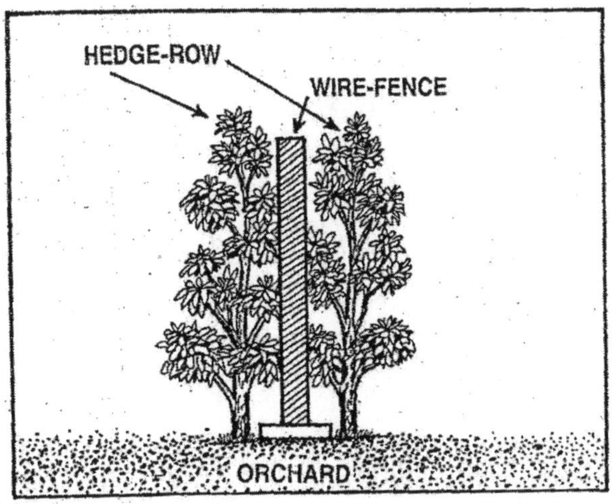

Figure 6.19: Hedge-Row Set at Either Side of Wire Fence.

Evidently, the advantages of live hedge as well as that of wire fence are derived by this type of combined method for which, the method is actually followed in many parts of India.

(b) Mud Wall and Hedge

Growing thorny hedge plants over mud walls or soil mounds (Figure 6.20) is a very common method that is maintained at the boundary of farms or orchards in many parts of India to prevent entry of trespassers and animals. Mud wall is made of low height and thorny hedges like *Agave, Opuntia*, dwarf ber, *Acacia* species *etc.* are usually planted for such purpose.

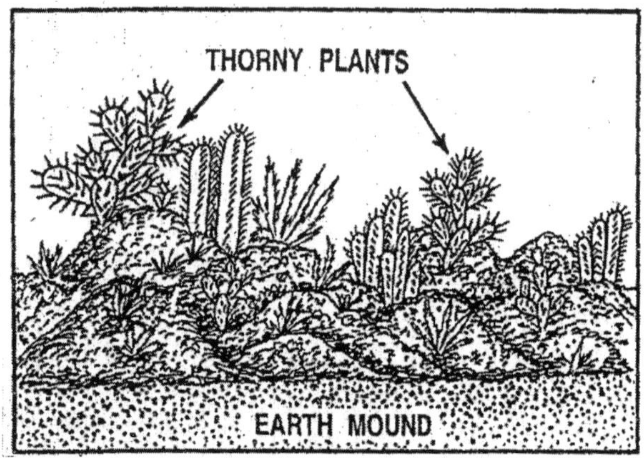

Figure 6.20: Thorny Hedge Planted on Low-heighted Mud Wall.

Beyond doubt, the method is extremely satisfactory against unwanted entry of human intruders, grazing animals as well as wild animals. Also that, it does not cost much and can be built up easily and quickly. In many parts of India, this method is largely followed and especially in drier tracts where xerophytic hedge plants are suitable materials.

(c) Brick Wall and Concrete Lattices

There is a preference by some people to construct at the boundary a brick wall upto a certain height and over that, the concrete lattices are cemented (Figure 6.21). The brick wall is usually made of relatively low height, *i.e.*, of 60-75 cm and over this, the lattices are set so that an ultimate standard height of about 1.5 metres is actually maintained.

A wall of this type has decorative value as well than when it is made up of bricks alone. The cost of construction is also perhaps lesser than the latter type.

(d) Brick Wall and Iron Railing

The method actually remains the same as has been described under the heading 6.4(3), with the difference that the iron railings are embedded at the top of the brick wall (Figure 6.22), instead of infixing them into the ground.

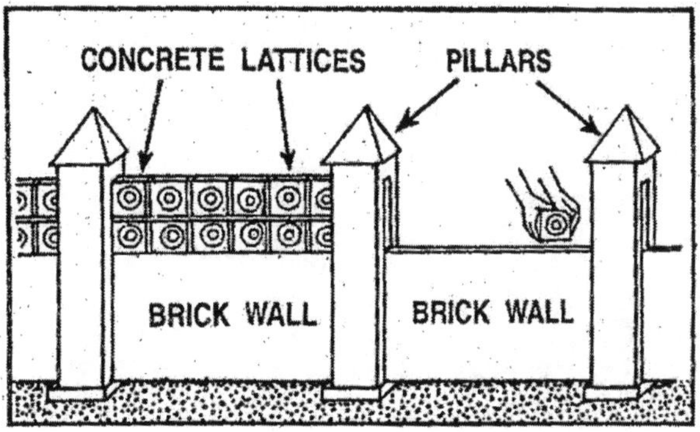

Figure 6.21: Concrete Lattices over Brick Wall.

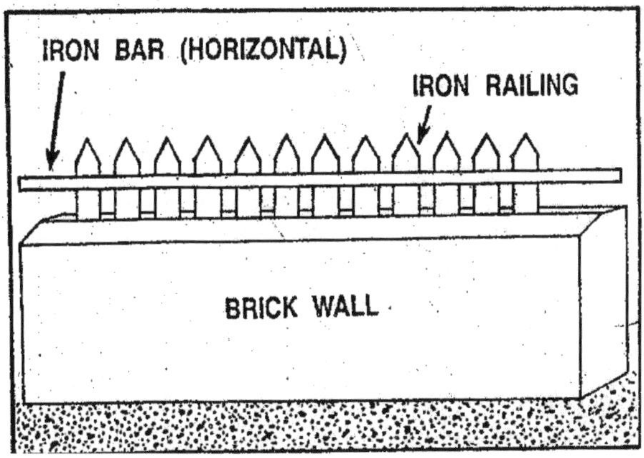

Figure 6.22: Iron railing over brick wall.

The brick wall is usually made of low height, *i.e.* 25-40 cm and at the top of this wall, iron railings are embedded upto the standard height that is desired. The combination makes the iron railings quite firm as compared to when these are infixed into the ground.

(e) Brick Wall and Wire Fence

Instead of iron railings, barbed wire fence may also be provided over brick wall. The brick wall may be of same height as in (d) and over this wall, the wire fence is to be installed. For better effect, the upper part of the wire fence may be projected inwards or outwards (Figure 6.23), or on both sides.

There may be many such devices where two or more methods are combined to frame a protective barrier at the boundary of the orchard and are adopted with success.

At the end, it is pointed out that there are orchrdists in many parts of India, according to whom adopting any extra measure to demarcate the orchard area is unnecessary, because it involves some expenditure and also that it has no utility. In fact, many of them have the argument that they have so nice and cordial relationship with the owners of the land surrounding theirs that any dispute with them is undreamt of. Such a co-operative spirit is undoubtedly most welcome.

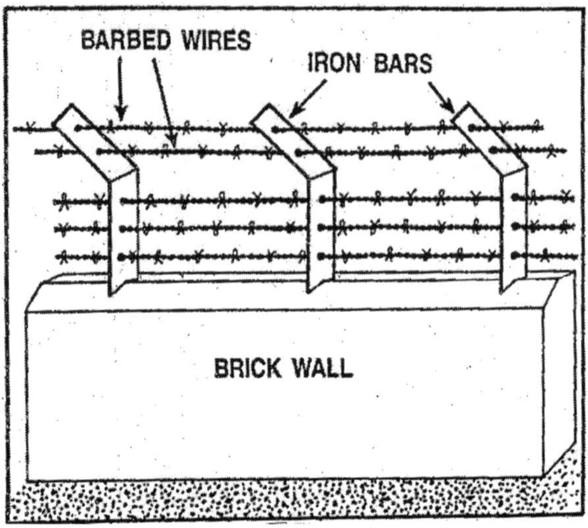

Figure 6.23: Barbed Wire Fence Over Wall.

However, it should also be remembered that cordial relationship between any two persons lasts long only when terms are so clearly defined that are easily understandable even by an unbiased third party. In unwelcome social dispute, which if arises on any day, it is only the previously set terms that we bring back to find solution. Accordingly, some initial expenditure to demarcate the orchard area should neither be set at defiance nor considered to be a prodigality. Easily recognizable and lawful demarcation of the orchard should be considered as a necessary step in orcharding and is not a wastage of money.

Chapter 7

Primary Operations of Land

Sometimes it is found that the land that has been obtained to set up an orchard has some *undesirable features* in it, which may create difficulty in successful culture of fruit crops. Various types of such shortcomings may be there, among which the following are more commonly noticeable in a land. Beyond doubt, the orchardist should get rid of these features at first, *i.e.*, before other operations are taken up in the land.

7.1 Removal of Unwanted Materials

1. Broken Parts of Buildings

Broken walls of buildings, hutments, *etc.* are some noticeable features in a land, obtained for cultivation or orcharding (Figure 7.1). It should be noted that concrete materials, mortars *etc.* are greatly harmful for growth of plants and hence, these materials need to be removed out completely from the orchard land. The underground structure, if any is present as foundation of a building should also be broken and the pieces brought outside the land, in order that the roots of the fruit trees to be grown there are not suffered.

2. Trees and Shrubs

Existence of one or more trees or shrubs growing here and there in the land is a very common sight (Figure 7.1).

If these tree(s) or shrub(s) remain present, they will compete with the young fruit plants for moisture and nutrients. As a result, proper growth of the fruit plants will not take place. Besides, the shade of those trees or shrubs will also fall on the fruit plants and for this, the latter will not have good growth.

Figure 7.1: Pre-existence of Stray Trees, Broken Buildings, Hutments *etc.* in the Land.

Therefore, it is necessary that these trees or shrubs should be cut back at first. It is also advisable that if such a tree is robust and a vigorously growing one, not only the above-ground parts but its main, thick roots (Fig 7.2) should also be exterminated as long as possible. This is because, in course of time, those thick roots would get decomposed and thereby, would form long tunnels in the underground part of the land. Due to formation of these tunnels, proper ramification and growth of the roots of fruit trees would be hampered. Besides this, the irrigation water that would be applied would be drained away through those tunnels to a greater depth or a long distance.

However, it needs mentioning in this connection that there is a class of people who strongly disfavour felling of any tree from anywhere. They are more staunch in this respect if the tree has any timber or other economic value and notably, if it has any religious significance. Such an attitude of a person which is based on a baseless sentiment hardly gets any support. This is in consideration to the fact that any stray tree has no justification to remain present in a land that has been obtained to set up an orchard as this will only hamper proper growth of the fruit trees for which the land is meant for. Nevertheless, any rare type of tree, especially of exceptional merit may gain support to be preserved and in such case, some area may be left for it without planting any fruit plant in that part. It is advisable that instead of maintaining the tree permanently there *in situ*, it should be propagated by suitable means and the propagated materials be planted elsewhere outside the orchard. Thus, when the propagated materials get established, the tree standing in the orchard land be felled.

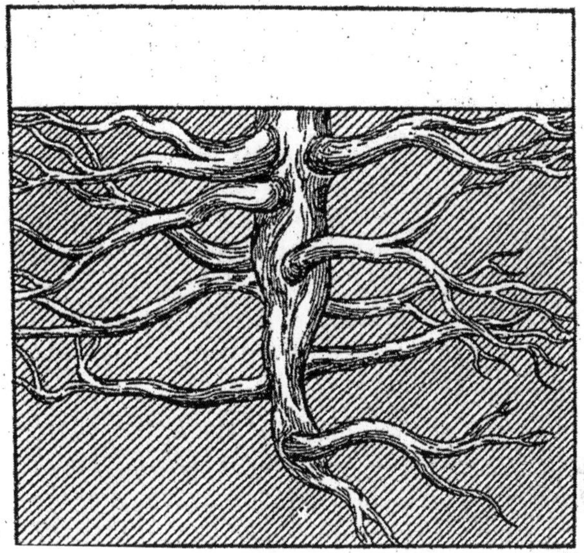

Figure 7.2: Thick, Anchorage Roots of Robust and Vigorously growing Stray Trees.

3. Long Grasses, Weeds and Stubbles

The land owned to set up the orchard may be found to grow weeds like ravenna grass (*Saccharum ravennae*) which is very tall and grows much in southern foothills of the Himalayas and north-eastern India, kans grass (*Cymbopogon citrates*), *Lasiurus scindicus* which grows much in Rajasthan, Mexican poppy (*Argemone mexicana*), *Cyperus rotundus, Pennisetum pedicellatum etc.* Existence of stubbles of crops is also very common in some land.

These must be removed after obtaining the land. *To* remove these materials, burning should never be done. It should be well-remembered that anything burnt directly on an agricultural land, a huge population of soil micro organism is destroyed, indispensable role of which is undeniable in cultivation of any crop. Decades may in fact, be required to replenish these micro flora and fauna in the land. Earthworms which remain present in the soil and are of great benefit in cropping performance are also lost. Hence, to remove these materials, mechanical method is considered best. Tractor drawn machinery such as flax-wing-cutter could be used which is very efficient in cutting even very tough vegetation.

7.2 Soil Work

1. Breaking of Clods

Particularly in heavy clay type of soil, existence of large clods in the land is very common. These are very hard but their underground parts may remain moist. Repeated beating them with bamboo pieces or iron rods should be done to break the clods. However, light irrigation or sprinkling of water in the land may sometimes become necessary to loosen the soil to facilitate breaking of clods. Tractor drawn rotavator or disc-plough or cultivator are efficient for the purpose.

2. Levelling

Before other works are taken up, it becomes necessary to level the land. Levelling of the land can be done easily with a long, broad and heavy piece of sawed timber. The wooden plank which is also known as patella is provided with hooks of iron at suitable positions. To operate, the hooks are tied with bullock or tractor drawn plough by means of strong rope. When the plough moves with the wooden plank, the land is suitably levelled with minimum shifting of soil. Levelling of land can also be done with concrete rollers. By working with these, large lumps of soil are also broken along with lavelling.

In hilly areas, flattening of the land may be difficult. In such soil condition, it is advisable to render the land into convenient gradations in consideration to the elevated and depressed pockets. The graded blocks are to be carefully made as faulty construction may lead to serious erosion of soil and disturb the evenness. It should also be noted that the graded blocks are to be built in their positions permanently and large modification of them is difficult to be made there.

3. Subsoiling

Subsoil is the layer just below the surface soil but above hard rock (hard pan). Subsoiling is an act of ploughing of the subsoil, *i.e.*, it is a type of *deep tillage*. It may be done of various depths but beneficial effect of subsoiling is hardly obtained unless it is done at least to 50-60 cm depth.

Many orchardists do not attach importance of this operation but an expert fruit culturist always advises this soil work before the fruit plants are set in the proposed orchard. There are many advantages of subsoiling, *e.g.*, it breaks hard packed soil layer, alleviates compaction of the soil, aerates the underground soil, destructs insects and other pests which live underground and so on. Subsoiling is most effective when the tillage is done with a *mould-board plough* which turns the soil while making furrows and ridges.

For the purpose of subsoiling, many types of mechanical subsoilers are used which are drawn by bullocks, tractors or mini-tractors. These consist of heavy vertical shanks, *i.e.*, standards which are mounted on a frame (toolbar) with shear bolts. Shanks are suitably curved and tips that could be replaced. A subsoiler should work upto 50 or 60 cm depth. Chisel plough are implements which unlike subsoilers work at relatively shallow depth may be used where deep suboiling is not necessary.

7.3 Control of Biotic Agents

The land obtained for the proposed orchard may even be infested with many types of animal pests. Although lesser infestations may be ignored, suitable control measures become necessary when they are in high abundance, in order to save the orchard plants to be set there and also the orchard workers. Evidently, this should be done before subsequent operations are taken up in the orchard. Among soil-inhabiting crop pests, rats, white ants, nematodes, rabbits *etc.* are of more common occurrence.

Rats are menacing problem in some fields and they result in death of young and mature plants, attacking their roots. Sometimes they climb on the trees also to eat fruits. Effective control of rats may be done by application of chemical rhodenticides. There are many such compounds available in the market and some of the common one are zinc phosphide, *e.g.*, Ratol, Commando, Zinctox *etc.* warfarin, *e.g.*, Rodafarin, Rotatox, Eoumafen *etc.*, bromadiolone, *e.g.*, Much, Ratan *etc.*, aluminium posphide, *e.g.*, Celphos *etc.*

Application of rhodenticides is done as baits, *i.e.*, they are mixed with some food which is liked by the rats and then put into the rat holes. Normally, 1 part of the chemical is mixed with 50 parts of food material and applied. For some rhodenticides however, mixing with food is not necessary as they are available in ready-made cake forms or tablets. Pumping in cyano-gas directly into the rat holes is also very effective but it should be done with extreme care. As regard cost, not more than Rs. 500 seems to be sufficient to control rats by rhodenticides in an hectare area of land.

Rats are also biologically controlled by snakes and hence, many agriculturists are not in favour of killing snakes, in order to control rats. However, although the snakes eat rats, they themselves are much more dangerous to the human beings and farm animals especially when venomous. While reducing the rat population, they increase their own population and thereby, their presence becomes a serious risk. Occasional sprinling of commercial phenol or even chlorinated water or powder and pouring the same in the holes where located is an effective method to repel anakes.

Termites and nematodes are other animal pests that are no doubt, of great danger to a fruit plant or any crop plant. If the orchard land is infested with these, they should be controlled before any operation is taken up. It should be remembered that presence of colony of white ants at any place may cause to infest large area in course of time.

Termites or white ants are of two types, which belong to the species (i) *Odonotermes redemanni, O. obseus* and *Coptotermes horni* which form mounds above the ground and (ii) *Odorntotermes assumthi* and *Microtermes obesi*, which dwell in the underground and they are more dangerous to plants. The eradication of termites is very difficult as they cover large area. The cyclodiene termiticides like Aldrin or Dieldrin are usually of common use to control this pest. Application of gamma-BHC of 5or 10 per cent is also used. The pesticides need to be thoroughly mixed in the soil. Other termiticides may be named as Chloropyriphhos 20 EC, or, Endosulfan 35 EC as Endocell 35 EC, Thiodan 35 EC, Sujan, Hildan, Thiodill, Ksulfan *etc.* The neem based biopesticide, Azadirachtin is also partly helpful.

To control nematodes, many of the methods that are done to sterilize the soil and described under the next Heading are found to be effective. For example, the chemicals that liberate toxic fumes like chloropicrin, vapam, mylone and formaldehyde are very good agents to kill nematodes. The gas given off by allyl alcohol is also quite effective. Other nematicies are carbofuran 3G as Anufuran, Furacarb 3G and mention may be made to the neem-based nematicides also like Nemazol-5 per cent, Achock, Rokshak, Margocide *etc.*

Apart from rats, termites and nematodes, some soil inhabiting inects also bring about great damage to the crops that are painted. To control most of the insects and larvae, gamma-BHC-2.5 per cent, Chlordane or Toxaphene or Aldrin-1.2 per cent are good. Besides, the chemicals that liberate toxic fumes as stated above are very satisfactory.

7.4 Soil Sterilization

Not only nematodes, protozoa, termites, earthworms, insects and many other animals, soil is the dwelling place for many *micro-organisms* as well. Some of these micro-organisms inhabit permanently in the soil while there are some, which colonize for short or long periods only according to availability of their food, *i.e.*, nutrients. Out of this large group of soil micro-organisms, there are some species which are *highly beneficial to the plants that are grown in the soil.* In fact, soil fertility and availability of nutrients are greatly influenced by these beneficial microbes. The addition of nitrogen to the roots of some species of plants by some species of the microbes through mutual rapport, termed in biology as *symbiosis* could indeed be said to be a classical example of plant sociology.

In contrariety, there also exists many micro-organisms in the soil which cause great harm to the plants. They are in fact, *pathogens* and thereby, cause diseases in plants. Since the causal organisms of these diseases inhabit in the soil, the diseases that are produced by the organisms are referred to as *soil-borne diseases* in phytopathology. These soil-borne pathogens bring about many devastating diseases in a number of agricultural crop plants. Needless to say, fruit crops are not spared from these diseases. In fact, soil-borne diseases in fruit crops have a long list and some of the common example are presented below.

The fruit tree, guava is threatened by a serious soil-borne disease known as *wilt* and every year, at least 5 per cent guava trees in India are totally killed due to this malady. The cause is the soil-dwelling fungal species, *viz., Fusariuum solanii* in most cases, although other fungi belonged to the species, *Macrophomina phaseolina, Cephalosporium etc.* are sometime responsible. Wilting assumes a serious problem in banana culture also, where it goes by the name as Panama disease and the causation is due to infection by the fungal pathogen, *Fusarium oxysporum f. cubense.* Many temperate zone fruit crops like strawberry, apricot, pistachio *etc.* sometimes become victim to wilt disease due to infection by the soil-inhabitinng fungal species, notably, *Verticillium albo-atrum* and *V. dahliae.*

Rotting of the collar region, *i.e.*, at the basal portion of trunk or roots are noticeable in many fruit trees. In papaya tree, such a disease deserves mentioning which goes by the name as collar rot, stem rot foot rot effectuated due to infection by the fungal species, *Pythium aphanidermatum*. Root rot is another soil-borne disease of this fruit tree due to infection by the same fungus or other fungal species like *Rhizoctonia solanii* or one or more species of the genus, *Fusarium*. Root rot disease is observable in some other fruit trees also, notably pear, caused by the fungus, *Phytophthora* species, in avocado, by the fungus, *P. cinnamomi*, in plum, walnut and some other temperate fruit trees by the fungal species, *Armillaria mellea* and so on.

A disease which is virulent in infant papaya seedlings in the nursery stage especially in poorly drained soil is damping-off. The fungal pathogens that cause the disease are in most cases, the same as are responsible for root rot disease. The citrus seedlings in the nursery are also suffered due to damping-off disease, the etiology of which is infection by the species of the fungi *Pythium* and *Rhizoctonia*. The crown gall tumours in apple deserves mentioning here, which is caused by the soil microbe, *Agrobacterium tumefaciens*, that is known to infect cherry, almond and some other temperate fruits as well. The above are only few of many other examples of soil-borne diseases encountered by fruit crops.

As regard the necessity of any action to be taken in controlling the pathogens in the soil before planting of fruit crops there, a section of orchardists is not in favour. They hold the view that it is not certain whether pathogens are existent in the soil and if so, whether they would infect the fruit crops, besides the fact that after disinfection, re-infection may also occur and above all, it involves some expenditure. The above points are no doubt, valid. Nevertheless, in scientific fruit culture, soil sterilization of the orchard land before planting as a routine practice has ample justification as an important sanitary measure than to skip out it. Accordingly, the general procedure of this operation has been laid down in the following.

First of all, it should be remembered that when sterilization of the soil is done, all the organisms that are existent in it are virtually killed, *i.e.* not only those which cause diseases but the beneficial ones also, although the objective is to destruct the harmful microbes only without hampering the helpful ones. Therefore, instead of the term, soil sterilization, *partial soil sterilization* is perhaps befitting to use. But it is pointed out that theoretically such a postulation is though bold, practically such a selective destruction of micro-organisms has great difficulty.

The different methods of soil sterilization practically used are of two categories, *i.e.*, physical and chemical and these are stated below.

1. Physical Methods

These methods aim at killing the organisms by heating the soil. Heating is done either in dry form or through steam, *i.e.*, in wet form. Dry heating is usually done by suitably placing electrically heated coils below the soil surface. For wet heating, steam is passed below soil surface at suitable pressure, which is generated in a boiler. Instead of using only steam, mixture of steam and air are usually used. A temperature of about 60° C for half an hour is considered satisfactory for destruction of most of the heat-resistant pathogens. It may be mentioned that the disease causing organisms are though killed, the helpful organisms survive at this temperature. For better result however, raising the temperature upto 80° C or even 90° C is suggested.

Heat sterilization has the greatest advantage in that, no chemical is used in this method and therefore, the soil is not chemically polluted, for which the crops grown are not harmed. But the method is cumbersome, costly and may be risky to the operators, for which it is not of much use especially for large-scale purpose.

Deep ploughing of the land and keeping it dry for few months with regular weeding is sometimes a practice to control pathogens present in the soil but it is a less effective method.

Another physical method known is flooding. This is particularly done in central America to get rid of the virulent Panama disease in banana which is cropped there extensively. The soil is flooded for several months before planting which causes destruction of the soil-borne fungus responsible for the disease.

2. Chemical Methods

In this method, the soil inhabiting pathogens are killed by the application of chemicals, which are either volatile or non-volatile. The volatile chemicals when applied, liberate fumes which are toxic to the micro-organisms. Thus, the fume kills the microbes and is given off into the air, such that nothing or very little part of the chemical remains present in the soil. Accordingly, the soil is not chemically polluted by application of these chemicals. The chemicals belonging to this group are usually referred to in plant pathology as *fumigants*. The non-volatile chemicals kill the micro-organisms on contact and they remain present in the soil for some period. Later, they are leached down by rain or irrigation water and some are decomposed slowly.

Among the volatile compounds, the following are usually much used (Ref. S.B. Chattopaddhaya. 2018. *Principles and Procedure of Plant Protection*. Oxford and IBH Publishing, N.Delhi).

(i) Chloropicrin

It is injected into the soil with the help of specially made injectors. Injection of 3-4 ml of the liquid is done at 10-15 cm depth of the soil at set points. Just after injection, the soil is lightly watered and as a result, the fume is liberated. In order that this emanated fume remains present in the soil in between the soil particles for some period and is not quickly dissipated into the atmosphere, after application of the chemical and watering, the soil should be kept covered with some materials like plastic sheets or even newspapers for at least 24 hours. It is expected that during this 24-hour period, the fume will kill all the microbes. The chemical is highly effective for many soil inhabiting pathogens that cause diseases like foot rot, root rot, wilt, damping-off *etc.* and also kills seeds of weeds and grasses that are present. To sterilize an area of land of 100 square metres, 45 litres of the chemical may be sufficient to use.

However, an important point should be remembered, that is, chloropicrin is a highly poisonous chemical. It is in fact, used by the police personnels to disperse the unruly mob. The fume causes burning sensation of the eyes of the people and hence, the police call it *tear gas*. But long period of exposure to this gas will not only bring tears from the eyes of the people but may also cause temporary blindness to them, besides damage to skin, nose, throat *etc.* The gas is more poisonous not only to the micro-organisms but also to the crop plants that would be grown there. Since the objective is to kill the micro-organisms in the soil and not the fruit crops or inter-crops to be grown, it is essential that when the microbes have been killed, the fume should be completely removed out from the soil. To do so after 24 hours of retention of the fume in the soil, *i.e.*, after removal of the plastic sheets or newspapers, the soil should be ploughed, or spaded or stirred with a hoe or a *khurpi.* On doing so, the soil should be left as such for at least 10 days. During this period of 10 days or more, all the fume present in the soil is expected to be removed out by dissipation into the air and only then planting of crops should be done. Sowing or planting before that may kill the crop plants completely.

(ii) Methyl Bromide

This gas is available in markets as an aerosol in metallic gas cylinders and also in liquid from. The active ingredient varies when available in liquid form. The method of application of methyl bromide and the precautions to be taken are same as in case of chloropicrin. The dose may be 10 kg of the compressed gas to sterilize an area of 100 square meters.

The rate penetration of methyl bromide into the soil is in fact, greater than chloropicrin and hence, kills microbes to a much greater depth of soil. But for some crops, notably onion, garlic, carnation etc. This gas is very muck toxic and if these plant are considered to be gowns as inter-crops in the orchard, soil treatment with methyl bromide is not advisable, for even traces of the chemical may kill those plants.

(iii) Mylone

This should be applied uniformly on the soil surface. To do so, this should be mixed with sand and evenly applied. After application, a spreader may be worked with as is used to spread the fertilizer on soil. To ensure uniform application, the chemical may also be made into a suspension with water and then after thorough stirring to be drenched on soil with a power sprayer or a watering can. A dose of 1-1.5 gram of active ingredient may be sufficient for an area of 100 square metres.

(iv) Vapam

Like mylone, this compound can be made into a suspension with water and drenched uniformly over soil surface through a power sprayer. The dose may be about 5 litres having 40 per cent active ingredient for an area of 100 square metres. The soil should be lightly irrigated after application. Unlike other chemicals where some toxic residues may be left in the soil, this does not the place by application of Vapam.

(v) Formaldehyde

This aldehyde chemical is highly soluble in water and is available in water soluble form, known as formalin which is a 40 per cent solution of formaldehyde. For soil sterilization, it is generally used as a 1 per cent solution and the dose may be 40-50 litres for 100 square metre area of land. The soil should be uniformly drenched with formalin and then covered for at least 24-hours as in case of Chloropicrin. Planting of fruit trees should not be done within 3 weeks of application.

(vi) Cresylic Acid

This acid decomposes to phenol and orthocresol which kill the disease causing microbes, especially the fungi. For treatment, uniform drenching should be done with a solution of 0.05 per cent as the active ingredient.

Among the non-volatile compounds, the commonly used ones are listed below.

(vii) Carbamates

These chemicals, *viz.*, Zineb, Meneb, Ziran *etc.* are applied on soil either as dust or in liquid form. In dust form, 250 gram of 15 per cent strength may be broadcast, *i.e.*, evenly applied for an area of 100 square metres and in liquid form, 40 litres of 0.05 per cent solution may be drenched on the same area of land.

(viii) Nitrobenzenes

These chemicals are also available as dust or in liquid form and the mode of application will be the same as in carbamtes. The active ingredient in dust form varies from 10-40 per cent and the dose may be 400-500 gram per square metre area. For spraying, upto 0.5 per cent solution may be used. The chemicals of this group are very much effective especially to kill soil inhabiting fungi, *e.g.*, *Sclerotium* and *Rhizoctonia*,, species which form sclerotia.

(ix) Copper Fungicides

These fungicidal chemicals should be sprayed at 15-25 percent concentration. The copper fungicides are particularly effective to control fungal species of the Class, *Phycomycetes.*

Chapter 8

Basal Application of Manure

8.1 Importance

When the orchard land is of large size, roads, paths and accessories become necessary to be provided inside it as have been stated under the Chapters 10 and 11. Application of manure after primary operation as stated in Chapter 7 is not necessary, because in such type of large-sized orchards, manures are to be applied directly in the cropping plots after these are laid out (Ref.: Chapter 12).

However, when the orchard size is not large, paths and accessories are usually not necessary to provide within it. In such type of orchards, the entire land is grown with fruit plants. Basal application of adequate amount of organic manure is of great benefit, despite the fact that manures are applied to the fruit plants during their cropping.

This application of organic manure is recommended primarily to improve the (i) *physical condition of the soil* for better cropping but it also adds some amount of (ii) *plant nutrients*.

8.2 Types of Manure to be Applied

Among the various types of bulky organic manures, compost, farmyard manure and green manure are more commonly used in India. However, nowadays, use of vermicompost is also increasing in many parts of the country.

1. Compost

Compost is made by decomposing plant materials. Any plant material is used for composting and sugarcane trash, paddy straw and various weeds are highly effective materials. As regard amount of compost that is necessary to be applied it may be stated that in an hectare area of land, 20-30 cartloads of compost (1 cartload = 4-5 quintals) is considered sufficient. Compost may be of (i) rural type (= farm compost), in which nitrogen (N_2), Phosphorus ($P_2 O_5$) and potassium (K_2O) contents are usually 0-4-0.8, 0.3-0.6 and 0.7-1.0 percent respectively and (ii) town compost which is made from street sweepings, dustbin refuse *etc.*, in which N_2, $P_2 O_5$ and K_2O contents are respectively 1.4, 1.0 and 1.4 percent.

According to compost specialists, the method which is normally practised to make compost in India has some faults. They suggested improved methods of composting such as *Indore method* and *Bangalore method.*

2. Farmyard Manure

This type of manure is made by decomposition of dung, litter and other materials left over from the fodder which is fed to the farm animals. It contains 0.5-1.0, 0.5-1.0 and 0.5-2.0 percent nitrogen (N_2), Phosphorus (P_2O_5) and potassium (K_2O) respectively.

3. Green Manure

The practice of green manuring refers to ploughing down of green leaves and succulent foliage part of plants in the soil in presence of copious moisture. On doing so, they are decomposed by the action of soil micro-organisms and thereby, are rendered to organic manure, besides providing other benefits as stated later.

There are in fact, several systems of green manuring that are practised as mentioned below.

(i) Seeds of leguminous plants belonging to any species are sown in the land and when the seedlings have grown to 30-60 cm height depending on plant species and are in succulent condition, the entire plants with root and branch are buried into the soil by ploughing deeply, or spading however, in presence of abundant moisture in the soil.

(ii) The same practice but sowing seeds of more than one species of leguminous plants.

(iii) Sowing seeds of non-leguminous plants and followed by the same practice.

(iv) Instead of ploughhing down the entire seedlings, the above-ground parts of them are cut off with a sickle or a large harvesting knife, which are left on the land and then turned into the soil by ploughing.

(v) Leaves and succulent twigs of herbs, shrubs or trees belonging to the species, *Gliricidia maculata, Erythrina indica, Ipomea carnea, Calotropis gigantia, Madhuca indica, Pongamia pinnata, Azadirachta indica etc.*, and above-ground parts of *mutha* grass and other weeds are evenly spread on the land and then ploughing down.

Among the different practices stated above, growing leguminous plants in the land and ploughing them render maximum benefit. This is because, apart from addition of organic matter, the roots of these plants act as host to some soil bacteria which take up free nitrogen from soil air and form nodules on the roots. On death of these bacteria, the nodules containing nitrogen remain adherent to the roots and when plouging of the green manuring plants is done and crop is grown, that crop gets that nitrogen from the nodules. It has been estimated that a leguminous crop produces 10 to 25 tonnes of green matter in an hectare area of land when ploughed and may also add 60 to 95 Kg of nitrogen in the soil. Besides adding nitrogen, some of the green manuring plants notably, green-gram, sunhemp, black-gram, cowpea *etc.* provide appreciable amount of phosphorus and potassium in the soil as well.

According to a report of the Indian Council of Agricultural Research, soil application of farmyard manure to the green manuring plants increases the root nodule numbers in leguminous plants besides increasing the cells of *Azotobacter* (micro-organism) in soil. In, some parts of West Bengal, in paddy cultivation, little amount of urea is applied to the green manure plants to encourage their quick and vigorous growth. (Some paddy growers in West Bengal even prefer to apply neem-coated urea each granule of urea in which has a coating of neem oil. This type of urea tends to slow down the dissolution rate of urea in the soil and thereby, enhances nitrogen availability for the next crop. However, neem-coated urea is rarely available now). Many paddy growers in West Bengal apply superphosphate which is required for the paddy crop, in the green manure plants. This is done as the clots of superphosphate are broken down and thus, the phosphorus is available readily to the paddy crop.

When green manuring is done with leguminous plants, the following aspects should however, be given due attention.

(i) Selection of the Species of Leguminous Plant

This is an important criterion because of the fact that suitability of the species is usually better in some particular type of edapho-climatic conditions. For example, *dhaincha* is very suitable in West Bengal, Bihar, the eastern States and Tamil Nadu. The plants of this species are also adaptable on alkaline and swampy land. *Guar* and also *berseem* are well-suited to the northern and central India, cowpea in Karnataka, lentil is largely used in the Kashmir valley and so on. It may be mentioned that, the plants of sunnhemp have wide adaptability and may be used in almost all over India. As regard supply of nutrients, nitrogen content is known to be higher in black-gram, sunnhemp, green-gram, cowpea *etc.* which should also be kept in view.

(ii) Time of Burying Down in Soil

The stage of growth of the plants when they are to be ploughed down is a very important consideration as bacterial decomposition is effectively and quickly performed when they are *succulent*. Lignification *i.e.*, hardening of the plants which results in advanced growth of them in fact, resists the decomposition by the soil microbes. When the growth becomes greater, the carbon matter in the plants also goes higher and nitrogen content goes lower and due to high carbon: nitrogen ratio, micro-organisms withdraw the nitrogen and other minerals, for which a nutrient deficiency may even result. The actual stage of ploughing depends upon the species of the legume grown and in general, a half-mature growth is considered satisfactory in most cases.

(iii) Adequate Soil Moisture Content.

Proper rotting, *i.e.* decomposition is not possible unless the soil contains copious amount of water. Due to some reason, if adequate water is not available in the soil when the plants are ready for ploughing down, it is rather advisable to cut back the plants with a sickle or a harvester and to leave the cut plants as such over the ground, instead of allowing them to grow further and becoming hard, *i.e.*, lignigied.

4. Vermicompost

This is a special type of organic manure which is found to give encouraging result in many agricultural and horticultural crops.

A report which has been provided by the Punjab State Council for Science and Technology states that vermicompost contains 1.6, 0.7, 0.8, 0.5 and 0.2 percent of nitrogen, phosphate, potassium, calcium and magnesium as compared to 0.5, 0.2, 0.5, 0.9 and 0.2 percent which are respectively obtained in farmyard manure. Vermicompost has also been found to contain 175.0, 96.5, 24.5 and 5.0 parts per million (ppm) respectively of iron, manganese, zinc and copper as against 146.5, 69.0, 14.5 and 2.8 parts per million in farmyard manure. The ratio of carbon and nitrogen in vermicompost has been stated to be 15.5 and in farmyard manure, 31.3.

The effectiveness of vermicompost depends on how it is prepared. Out of about 3000 species of earthworms, 509 species are found in India. Out of these, the species belonging to the *epizoic group* are considered ideal. Superior quality of vermicompost should contain 15-25percent moisture, its colour should be brown to black and bulk density should be 0.7-0.9 gram per square centimeter.

Application of compost, farmyard manure or vermicompost must be followed by their thorough incorporation into the soil by ploughing and cross ploughhing. The land should be levelled after that. If these organic manures do not contain copious moisture or the soil is not wet to workable condition, water may be sprinkled over them. Instead of that, very light irrigation in a scattered manner may even be applied at some places.

Chapter 9

Windbreak Rows

9.1 Role of Wind on Plant Life and Fruit Plants

No higher plant or animal can live without direct exposure to air and fruit plants are no exception to that. However, blowing air *i.e., wind can render great benefit or serious hazard to a fruit plant* according to the intensity of its blowing. Principal benefits or harmful effects of wind may be stated as follows:

1. Beneficial Effect

Blowing air in slow motion in the inside part of a fruit plant brings fresh air by drifting away carbon dioxide, ethylene or other gases that accumulate there as metabolic end-products of the plant organs. High relative humidity in the interior part of the fruit plant that forms due to evaporation and transpiration from the leaves is also lowered by wind. Besides, fresh air drains out *cold air* which sometimes gets trapped in the inner part especially in trees of dense foliage or settles down under it in the winter season.

The above are the good effects of wind to a plant which cause to increase its rate of photosynthesis, respiration, uptake of water and nutrients. Eventually, these results in higher metabolic activity of the fruit plant and as a result, its growth and developmental activities continue in an undisturbed manner. The ultimate result is, productivity of the fruit plant is increased, both quantitatively and qualitatively.

2. Hazardous Effect

If the wind blows at a high velocity, it makes harm to a fruit tree than any good. Evidently, the harm is greater with higher the velocity of wind and more so, if the wind is hot. In fact, in many parts of India, the orchardists experience that high wind is perhaps the *most serious threat* in fruit culture. The hazardous effect of wind exists in many other parts also, though not in very serious form. As mentioned earlier under the Heading, 3.1 (8) in the Chapter 3, some parts of India are in fact, much prone to cyclonic storm in almost every year and in those geographical belts, the ravages are more virulent.

It may be stated that almost in every year, near the end of summer and advent of south-west monsoon, the coastal States of India are threatened by cyclonic storms, known as *tropical cyclones*. During this season, 5 to 10 storms actually originate in the Bay of Bengal and the Arabian sea. These cyclones bring about torrential rain, storm surges, gales and squall and above all, loss of human lives and property to a great extent. There is a record of deadliest tropical cyclone in eastern India during the month of November in the year, 1970.

Beyond doubt, trees are not spared from the destructive effect of high wind, storm or cyclone. In respect of fruit trees, it needs pointing out that on an average, at least 5 per cent of them in India are estimated to be badly damaged or even rooted out every year due to storm.

The fruit trees are harmed by high wind in a number of ways and the important ones may be listed as follows.

(i) Breaking or blowing down of branches.

(ii) Tilting.

(iii) Tearing out with root and branch, *i.e.*, extermination, especially of shallow-rooted fruit trees like papaya, sapota, banana, pomegranate, fig, custard apple, phalsa *etc.*

(iv) Falling of flowers, fruits and leaves.

(v) Cracking or shriveling of many fruits.

(vi) Drying of stigmatic surface of flowers and thereby, disallowing pollination.

(vii) Covering stigmatic surface of flowers with dust, hampering pollination and also on leaves, impairing photosynthesis, respiration and transpiration.

(viii) Exerting stress to the activity of the pollenizer insects.

(ix) Increasing the rate of evapo-transpiration from leaves and drying of shoots.

(x) Loss of soil moisture due to high rate of evaporation.

(xi) Blowing away of the fallen leaves from the orchard land that could otherwise improve it in adding biomass.

(xii) Blowing away of surface soil from the orchard land which is rich in fertility.

(xiii) Drifting of sand over the orchard land, if any dune or desert exists in the vicinity.

Thus, wind blowing at high velocity in a place (may though be a powerful source of natural energy and harnessed for generating electricity or other beneficial use) is a jeopardy in the culture of fruit crops in that place.

9.2 Windbreak Rows of Trees and Shrubs

The above discussion therefore, makes it imperative that especially in areas that are threatened by high wind, suitable measure is necessary to be taken to protect the orchard trees from the destructive effect of this calamity, if fruit culture is desired to be done as a successful farming enterprise. This is accomplished in an orchard by the setting up of what is known as windbreak rows of trees and shrubs.

As a matter of fact, any device which reduces the velocity of wind is referred to as a windbreak. Thus, a brick wall which is able to offer resistance to the wind is also a windbreak and one may consider to construct such a wall along the orchard boundary for protecting the fruit trees grown there. However, such a proposition gets little support, because construction of such a wall not only involve very high cost but also it is a fact that a brick wall is liable to be broken easily if the velocity of wind goes very high.

In fruit culture, an effective contrivance from the practical point of view would be to set up one or more rows of tall and vigorously growing trees and shrubs at the boundary of the orchard to serve as windbreak. The row should evidently be maintained at right angle to the direction of the wind force. To secure better effect, compactness should be maintained by planting the trees or shrubs closely together in the row. The effectiveness will be greater if more than one row instead of one is set up and even with inclusion of a row of tall and bushy grasses. These rows of vegetation are known as windbreak rows of trees or shrubs.

As the function of windbreak rows of trees or shrubs is to offer resistance to the wind of high velocity, they should be very hardy in nature to such an extent that they do not break easily by high wind. They should also be dense and bushy in appearance so as to perform the act of an effective barrier to the high wind. Above all, the total height that would attain by the trees or shrubs in the rows on maturity should be greater than the orchard trees, for in such case, they are capable to protect the orchard trees or shrubs against high wind, which blows mostly at the upper layer of atmosphere.

9.3 Necessity of Windbreak Row

The windbreak row(s) of trees or shrubs offers resistance to the wind and by that, it reduces the *velocity* of wind. At the same time, the row(s) changes the direction of the wind, such that it tends to move upwards. In that way, the windbreak row protects the orchard trees along with the orchard soil against possible damage due to wind storm. It is usually observed that a straight wind with a velocity of 48 km per hour, when strikes against an effective windbreak row of 10.5 metre height reduces wind velocity to 16 km per hour for 45 metres and to 24 km per hour for

60 metres on the leeward side (Leeward: The side turned away from the wind. Windward: Towards the direction from which the wind blows).

How much effective the windbreak row(s) is, would much depend on a number of factors. These are, species of the trees or shrubs that are planted, location of the rows and their numbers laid out, planting density, management practices and so on. Evidently, benefit of the windbreak row(s) are derived accordingly in regard to resistance against wind.

However, apart from protecting the orchard against damage due to high wind, some species of windbreak trees and shrubs render other benefits also. Thus, if fruit trees, *e.g.*, mango, jamun, jackfruit, hog-plum, khirni, sapota, palmyra-palm, kendu, ber *etc.* are planted as windbreak row(s), some fruits are also obtained from them. It should be remembered that if fruit trees are planted to serve as windbreak rows, the trees produced from seeds should be used as they would be hardy, although they might not produce fruits of good quality.

Similarly, if timber trees are planted as windbreak, their branches may be occasionally pruned for use as timber. Leaves of *Cassia siamea, Ziziphus* species, *Inga dulcis etc.*, and pods of *Acacia arabica, Parkinsonia aculeata, Prosopis juliflora etc.* are used as fodder. If mulberry is used as windbreak tree, it would be useful in rearing silkworms as well. Likewise, planting of bamboo ensures economic benefit apart from their value as windbreak.

However, despite the above beneficial effects of windbreak rows, the following demerits of them may also be cited.

(i) The windbreak rows occupy some land in the orchard and this may be an economic loss particularly when the land has been purchased at high price.

(ii) The moisture and nutrients from the orchard soil may be drawn out by the roots of the windbreak trees and thereby, the fruit trees grown in the orchard may be deprived. As a matter of fact, the windbreak trees being mostly wild in nature send forth their roots to long distances in order to seek moisture and nutrients. These roots may also come to the orchard soil and due to wild nature, they ramify quickly and vigorously when get fertile soil in the orchard. In competition with roots of the fruit trees, they win in respect of withdrawal of moisture and nutrients from the orchard soil and accordingly, cause the roots of the fruit trees to suffer. (This problem could be greatly averted by adopting the device described later under Chapter 9:8)

(iii) Shade of the windbreak trees may fall over the fruit trees in the orchard and thereby, hamper growth and productivity of the latter.

(iv) The windbreak trees and shrubs harbour many insects or animal pests and disease organisms which may invade the orchard trees also. When fruit trees are planted in windbreak rows, they serve as roosting places for birds which may damage the fruits of the orchard trees as well.

From the above discussion, it would be apparent that establishment of windbreak row or rows in fruit culture has its merits, though some demerits are also associated with it. Therefore, it may be a confusing situation for an would-be orchardist to decide whether the windbreak rows should or should not be set up at the boundary of the proposed orchard.

The satisfactory answer to this question is, windbreak row has a dire necessity in an area which is regularly threatened by high wind to such an extent that ravages the trees in that area. Despite the demerits associated with windbreak rows of trees or shrubs, provision of them is highly justifiable in that area and should be regarded to be a must, if successful fruit culture is desired. The number of rows of windbreak, their locations and the species of trees or shrubs that should be selected is again, a pertinent question and this should be decided in consideration to the factors like virulence of wind in that locality, edapho-climatic situation, species of fruit crops grown, size of the orchard and so on. To the other extreme, provision of windbreak row does not appear to have its essentiality in a locality where the problem of high wind causing heavy damage to any tree has virtually been unknown or have no significance. Nevertheless, it needs mentioning that even in such situation having practically no problem of wind or storm, at least a single row of windbreak may prove its worth.

9.4 Location for Setting up Windbreak Row(s)

The objective of the windbreak row(s) is to protect the orchard trees against damaging effect of high wind. Hence, it should be allocated from where the wind enters into the orchard, *i.e.*, at the boundary of the orchard. But wind may enter into the orchard from more than one direction. Hence, it may become necessary to set up the windbreak rows at more than one boundary of the orchard. Another important point is, if the size of the orchard is very large, it may even be necessary to set up windbreak rows not only at the boundary(ies) but even in the inside of the orchard also.

These have been elaborated in the following.

1. Only at the Boundary(ies)

When the windbreak row(s) would be set up only at the boundary(ies) of the orchard, it should be allocated to align at *right angle* to the direction of the wind force and that, it should be allocated at that side of the orchard from which side the wind blows inside. This is shown in Figure 9.1.

In some areas, it is experienced that wind of high velocity *comes from two directions at different seasons*. For protecting the orchard trees in these localities, it is advisable to set up windbreak row(s) at two boundaries of the orchard, instead of one boundary only as shown in Figure 9.2.

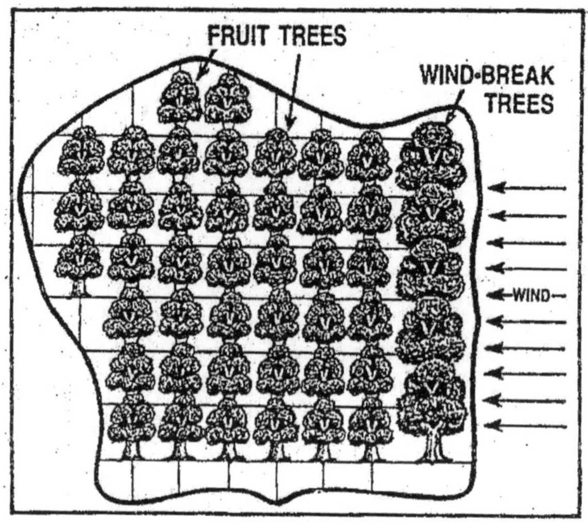

Figure 9.1: Windbreak Row Set at One Boundary of the Orchard.

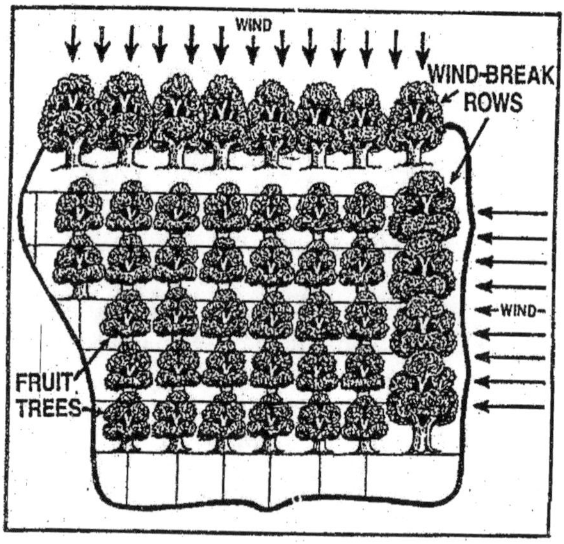

Figure 9.2: Windbreak Rows Set at Two Boundaries of the Orchard.

2. Both at the Boundary(ies) and in the Inside of the Orchard

If the lengthwise span of the orchard is considerably large, there may be necessity to set up windbreak row(s) not only at the boundary(ies) of the orchard but even inside the orchard also.

It may be pointed out that when the turbulent windstorm strikes the windbreak row, its *direction is at first changed upwards*. But on moving at the upward direction upto some distance, it will again tend to move downwards like an arch which has been shown in Figure 9.3.

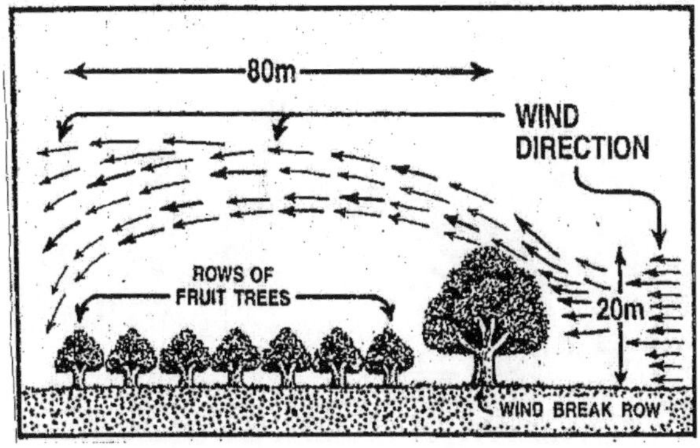

Figure 9.3: Wind is Directed Upwards on Striking at the Windbreak Row and again moves Downwards.

It is generally estimated that blowing *upto a distance of about four times* the height of the windbreak trees, the wind again changes its direction and tends to blow downward. Thus, the fruit trees in the orchard which are located at a distance of four times the height of the windbreak row(s) will be protected but those fruit trees that are located beyond this distance will be struck by the wind as it has then moved downwards.

For example, if the height of the windbreak row at the boundary is 20 metres, it is capable to protect the fruit trees located in the orchard upto a distance of 20 multiplied by 4, *i.e.*, 80 metres measuring from the windbreak row (Fig. 9.3). But if the length of the orchard is much greater than 80 metres, a windbreak of 20 metres in height is not considered enough. In such case, as shown in Figure 9.4, a *secondary row of windbreak* becomes necessary to be planted within the orchard land and this windbreak row should be set parallelly to the row(s) which has been set at the boundary. Thus, if the windbreak row(s) at the boundary has a height of 20 metres and the lengthwise span of the orchard is much greater than 4 times of it, in other words, more than 80 metres, the secondary windbreak row should be set at a distance of 80 metres from the boundary windbreak row.

It should however, be noted that by establishment of a windbreak row within the orchard, some part of the valuable land in the orchard remains unproductive. Therefore, planting of a windbreak row inside the orchard land should be done only when the wind comes in virulence. That is, windbreak row inside the orchard may be provided, if it is considered essential to protect the fruit trees located at the distant part of the orchard against *definite damage by wind* and not merely, in considering the lengthwise span of the orchard.

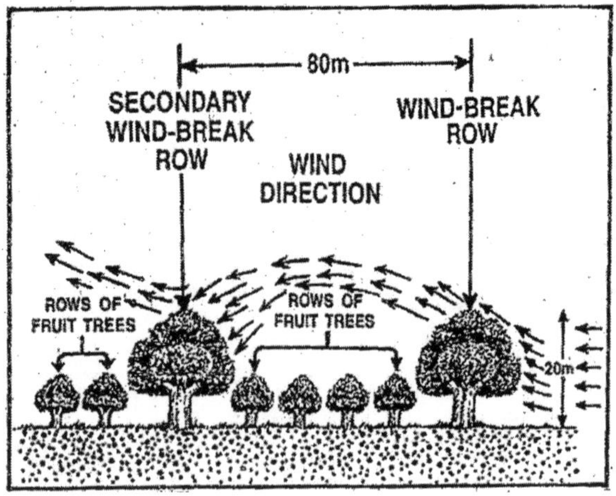

Figure 9.4: Windbreak Rows Planted at the Boundary as well as Inside the Orchard.

9.5 Number of Windbreak Rows to be Set up

In consideration to necessity, the orchardist should decide whether to plant only a single row of windbreak, or more than one row. In the latter case, the rows should be spaced closely together.

Under Indian condition, setting up of a single row of windbreak at the boundary of the orchard is a common practice. But it should be remembered that in parts of the country where wind current is severe, a single row may not be strong enough to afford adequate resistance to wind. The violent storm may even cause breakage of the branches of the windbreak trees or shrubs. Besides, in single row of windbreak, wind may enter into the orchard through the lower side, especially when it is not bushy in appearance.

To avoid such difficulty, two or three rows or even more than that may be necessary to be planted. In this system of setting up of multiple rows, selection of the species is however, an important consideration. In other words, those species of trees or shrubs are to be selected for planting which attain different heights when they are in grown up condition. Care should be taken in planting such that the row of dwarf trees or shrubs is set at the outer side of the orchard, *i.e.*, at the *windward side* and after that row, the relatively tall ones at the inner side to form as the intermediate row and finally, the row of still taller trees that should be set at the innermost side of the orchard, *i.e.*, at the *leeward side*.

By panting in the above manner, a gradient would accordingly be maintained, such that the windbreak row of the dwarf trees would be positioned at the outer side of the orchard and would be followed by the relatively tall trees at the inner side of it (Figure 9.5). The trees or shrubs in each row should however, be closely spaced to form a *bushy and compact appearance*.

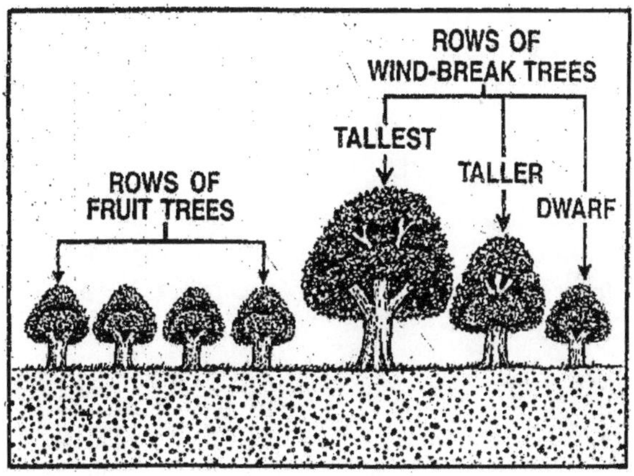

Figure 9.5: Three Adjacent Rows of Windbreak Trees with Dwarf Ones at the Outer and Taller Ones at the Inner Side.

The multiple rows of windbreak trees or shrubs would although occupy some extra land but they prove to be much more efficient in resisting strong wind and which is difficult if a single row is maintained. It is apparent that the vegetation formed by the windbreak trees and shrubs covers a large space from low to high level. Accordingly, the wind that strikes at the lower level gets resistance by the row of dwarf trees or shrubs and the wind that blows at higher level is resisted by the rows of the taller trees. It may also be noted that any possible damage to the windbreak trees or shrubs that may result due to strong wind is also less as the wind current is shared by the trees or shrubs that exist in all the rows.

By establishing multiple rows of windbreak, some valuable land of the orchard no doubt, does not give return. Therefore, this system should be adopted only when the actual necessity is felt and not everywhere as a rule. That is, the decision to maintain multiple rows should be taken in consideration to the factors like severity of the wind current in the concerned locality, species of fruit crops to be grown and their cultural methods to be adopted, species of windbreak materials to be selected and so on.

9.6 Suitability of Plant Species

Although trees are more commonly used, many shrubs are also found to be much suitable for planting as windbreak in some places. Again, some species of grasses which are tall and highly bushy in appearance may serve as excellent materials when combined with rows of trees or shrubs. In fact, the efficiency of a windbreak row depends mostly on the plant species that have been selected for planting.

In selecting the plant species to set in windbreak row, some aspects should be given adequate attention. It should be noted that the windbreak trees or shrubs aim at affording resistance to the strong wind and hence, they should be structurally hardy to withstand the wind force without felling, tilting or breaking branches by it easily. In this respect, species of trees or shrubs which are of wild nature are more effective as they are hardier and develop strongly built roots systems. If fruit trees are used *seedlings should be preferred as they are more hardy* than the vegetatively propagated ones. Resistance against pests and diseases should also be taken into consideration.

Quite a large number of plant species can be named which are sustainable for planting as windbreak. For back in 1958, Kaith and Erasmus (Ref. : Kaith, D.C. and Erasmus, I. 1958.

Wind-breaks and Shelter-belts

Wind-breaks and *Shelter-belts*. Farm Bulletin No.46, Indian Council of Agricultural Research, New Delhi) authored the bulletin which is a highly valuable guide that may be paid emphasis in regard to planting of various species of windbreak trees and shrubs in several zones of India. From their list as presented below, species of plants to be used for planting in particular location should be judiciously selected.

1. Northern Region

First Row (Windward side)

Euphorbia royleana, E. neriifolia, Saccharum munja, Arundo donax, Sesbania aegyptiaca, Lawsonia alba, Agave americana, Thevetia neriifolia, Tecoma stans, Ipomea carnea, Vitex negundo, Zizyphus nummularia, Acacia jacquemontii, Calotropis procera and *Capparis aphylla.*

Second Row (Middle)

Zizyphs jujube, Parknsonia aculeata, Prosopis spicigera, P. juliflora and *Tararix articulata.*

Third Row (Leeward side)

Acacia arabica, Albizzia lebbek, Dalbergia sissoo, Azadirachta indica, Eugenia jambolana and *Cedrela toona.*

2. Rajasthan Desert Area

First and Second Rows (Windward side)

Saccharum munja,Calligonum polygonoides, Leptadenia spartium, Panicum antidotale, Cenchrus ciliaris, Zizyphus mularia, Cappartis aphylla, Tamarix articulata, Pennisetum typhoidenum, Balanites roxburghii and *Kochi indica.*

Second and Third Rows (Leeward side)

Parkinsonia aculeata, Prosopis spicigera, P. juliflora, Tecoma undulata, Acacia arabica, A. leucophloea, A. senegal and *Ricinus communis.*

3. Central Region

First Row (Windward side)

The species recommended for the northern region are *Pithecolobium dulce, Sesbania grandiflora, Agave sisalana, Luecanea glauca, Jatropha curcas, Gliricidia maculata* and *Ricinus communis.*

Second Row (Middle)

Acacia arabica, A. catechu, Melia azedarach, Pongamia glabra, Cassia siamea and *Boswellia serrata.*

Third Row (Leeward side)

Tectona grandis, Mangifera indica, Dalbergia latifolia, Tamarindus indica, Hardwickia binata, Terminalia arjuna and *T. tomentosa.*

4 Southern Region

First Row (Windward side)

Euphorbia tirucalli, Gliricidia maculata, Agave species, *Sesbania aegyptiaca, Jatropha curcas, Tephrosia canadida, Casuarina equisetifolia, Erythrina indica* and *Bombusa* species.

Second Row (Middle)

Anacardium occidentale, Acacia moniliformis, A. mollissima, Prosopis juliflora and *Bambusa* species.

Third Row (Leeward side)

Eucalyptus species, Acacia arabica, Tamarindus indica, Albizzia lebbek, Grevillea robusta and *Borassus flabellifer.*

5. Eastern Region

First Row (Windward side)

Agave sisalana, Arundo donax, Vitex negundo, Imperata cylindrica, Tephrosia candida, Lennea grandis, Bambusa species and *Musa* species.

Second Row (Middle)

Casuarina equisetifolia and *Anacardium occidentale.*

Third Row (Leeward side)

Coconut palms, arecanut palms, *Shorea robusta, Dalbergia sissoo, Mangifera indica, Eugenia jambolana, Madhuca lalifolia* and *Artocarpus heterophyllus.*

6. Sea Coast Region

First and Second Rows

Spinifex squarrosus, Panicum turgidum, Ipomea biloba, Pongamia pinnata, Indigofera paucifolia, I. aspalatoides, Euphorbia tirucalli, Tamarix articulata, and *Erythrina indica.*

Second and Third Rows

Tamarix articulata, Prosopis juliflora, Casuarina equisetifolia, Anacardium occidentale, coconut palms, *Poinciana alata, Ficus* species, *Borassus flabellifer* and *Thespesia populnea*

9.7 Propagation Methods to be Adopted

1. Use of Seed

Where propagation by using seeds is possible, it is no doubt, the best method for raising windbreak trees or shrubs. The reasons are stated below.

(i) It is by far, the easiest of all propagation methods and does not require any extra skill.

(ii) Trees or shrubs coming out from seeds are *hardier.* In other words, they have greater ability to withstand adverse edapho-climatic situations than those which are propagated by vegetative means.

(iii) Seedling trees or shrubs are usually vigorous in growth and assume tall and spreading appearance in mature condition.

(iv) They develop highly spreading and strongly built root systems.

(v) Above all, seed propagation is least expensive as well.

Despite the above merits, it needs pointing out that there are some seeds where problem is encountered in germination and thus, raising seedlings easily from them becomes very difficult. The difficulty in germination may arise if the seeds are not viable, or are not fully mature, or attacked by pests or diseases, or are not properly nourished and so on. But even when these conditions are ruled out, the seeds may not respond to germination easily, although when they are subjected to optimum environmental conditions necessary for germination.

The failure to easy germination for such seeds is due to germination blocks already present in them and the condition is known by the general term as *dormancy* of seeds. The dormancy of seeds is however, a broad term in that, it is a complex situation and the causes of it may be either physical or physiological or both. The physical block of germination is usually imposed by hard and impermeable seed-coat or seed-covering. (It should be noted that seed-coat and seed-covering are not same. Seed-coat is a botanical term and refers to the *testa* and *tegmen* of the seed. Seed-covering is *a term used in horticulture* to denote both seed-coat as well as any other organ, which is botanically not a part of the seed but remains attached over the seed so tightly that it is difficult to separate it from the seed and hence, seeds are sown along with these coverings attached to them, for example endocarp in the seeds of mango, peach, plum *etc.*, which in fact, is a part of the fruit and not seed).

Seed-coat or seed-covering sometimes acts as a mechanical barrier in exchanges of gases with the inner parts of the seed, *i.e.*, the embryo and the endosperm, supply of moisture to these inner parts, or emerging out of the radicle and the plumule from the seed. On the other hand, the physiological block of germination is due to the natural existence of some chemicals which cause to inhibit or retard germination. These native chemical substances, *i.e.* the metabolic products are mostly hormonal in their action.

To overcome the germination blocks, some pre-treatment of seeds are however, found to be very effective steps. For example, when the hard seed-coat, or the seed-covering, *e.g.,* persistent endocarp of the fruit is the actual cause for non-germination, this can be loosened much by damaging it with a device, known as *scarification*. The scarification may be done by mechanical means, such as rubbing the seeds with sand, rubbing between two sand papers, rubbing with mortars, or stone chips *etc.*, or by chemical means which is done by quick or short time dipping in concentrated sulphuric acid or in other corrosive chemicals. These treatments scarify the seed-coats or the seed=coverings and accordingly, make them permeable to moisture and respiratory gases. Thus, the difficulty faced by the inner part of the seed, *i.e.*, the embryo and endosperm in getting the requirements for germination is overcome and this results in protrusion of the germinating embryo without difficultly.

Soaking the seeds in clean running water for a considerable period is also a very effective method. This practice not only loosens the seed-coats (or the seed-coverings) but also leaches out to a great extent the native chemicals that cause germination inhibition. If the facility for soaking in running water is not available, soaking may be done in still water but the water should be changed at interval of few hours. Another effective method to overcome dormancy of seeds for many species is *moist-chilling*, which *consists of soaking the seeds* to become hydrated at first and then exposing them to low temperature. The degree of hydration, temperature and duration of exposure of the seeds would however, vary according to species.

Lastly, it may be stated that some hormonal or non-hormonal chemicals are available which have proven ability to ease germination to a remarkable extent. Among many such chemicals of this group, the commonly used ones are gibberellic acid, ethephon, auxin, potassium nitrate, thiourea *etc.* 'Germitone', a commercial product is also highly effective to overcome dormancy and facilitate quick germination of seeds of a large number of species. For treatment with any physical or chemical agent to promote germination, it should be kept in mind that the strength and duration of treatment should be properly determined for the type of seeds and should not be considered same for seeds of all species.

To raise the windbreak plants, sowing the seeds directly *in situ* is also a very good method. In this method, few seeds should be sown closely together at one place, instead of sowing only one seed at the place where the row is made. When the seedlings would emerge out from them, only those which are vigorously growing should be retained and rest pulled out with roots. The same process should be repeated two or three times at an interval of a week or so, such that ultimately only one seedling is left to grow. To induce germination of seeds and subsequent growth of the seedlings, regular watering, weeding and culture at the row are no doubt, the indispensable steps. The method has the advantage in that, transplanting of the seedlings is avoided.

However, if *in situ* method of sowing seeds is avoided, seedlings have to be raised in the nursery at first and then these are to be transplanted in the row. For this purpose, it is a better practice to dig sunken bed of about 20 cm depth in a semi-shade situation and fill the bed with a mixture of well-rotten farmyard manure, leaf-mould, garden soil and sand in suitable proportions. Seeds should be sown at desired depth and spacing and the bed is regularly watered lightly and frequently. When the seedlings would attain maturity for transplantation, they should be carefully uprooted without damaging roots from the pre-moistened bed and transplanted in the row.

In the above method, there is chance of destruction of many roots of the seedlings during their uprooting. To avoid this, a better method may be to raise the seedlings in small polythene sleeves. To do this, ordinary thin-gauged polythene tube (ribbon) of 6-10 cm diameter should be taken and this should be cut into 10-16 long pieces. Each of these pieces having both ends opened is filled with a mixture of sand, garden soil, leaf-mould and manure and is placed vertically over a layer of sand and soil mixture which is raised to few centimeters from the ground (Figure 9.6)

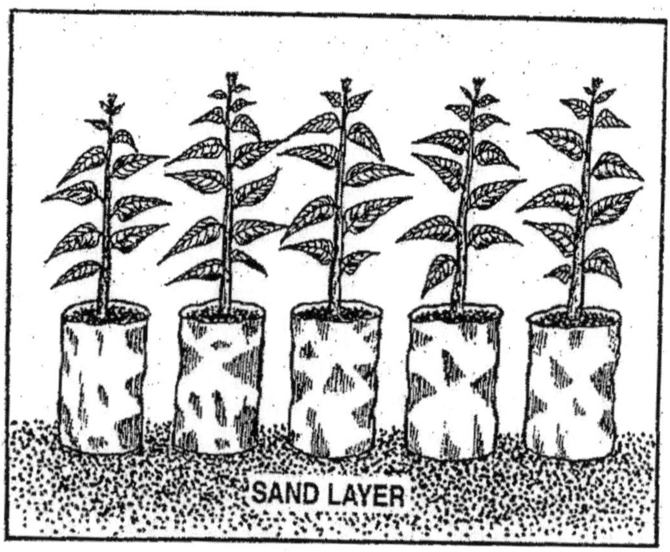

Figure 9.6: Seedlings Raised in Polythene Sleeves.

In each of the filled-in sleeve, 2-4 seeds are sown at desired depth and it is regularly watered. Optimum moisture for germination is obviously maintained in the medium of the sleeve as the excess water would freely drain out from the bottom. When the resulting seedlings would be few centimeters tall, 2-3 of them which are more vigorous should be retained and the rest are uprooted. Again, after a few days of their growth, only that seedling which is most vigorously growing should be retained and the rest are rooted out, such that each sleeve would ultimately maintain only one seedling. The sleeves with the medium containing seedlings should be watered and maintained till they become ready for transplanting.

For transplanting, the sleeve containing the seedling grown in the medium is brought to the respective position. Where planting is to be done, the soil is scooped to a little depth. Then the sleeve is lengthwise cut with an incision given by a razor blade, the polythene is removed out and the seedling along with the medium is carefully placed in the scoop dug out in the soil. After transplanting, the medium should be firmly pressed and regularly watered. This method has the merit in that, the roots of seedling which are the absorbing organs of it are not damaged if the operation is performed with care. Water management of the seedlings is also efficient in this method such that neither moisture deficit nor poor drainage takes place in the medium of the sleeves.

2. Use of Rooted Plant Materials

Apart from seedlings, cuttings, *i.e.*, rooted stem-cuttings or layerlings, *i.e.*, rooted layers are sometimes used as propagating materials in some species. Protrusion of adventitious roots takes place easily in easy-to-root species but this may be difficult in difficult-to-root species. In the latter case, use of auxin alone, or auxin in combination with phenolic compounds may be of help to induce roots. Allowing root formation of leafy stem-cuttings in *mist-house* may also give good result. For some stem-cuttings, application of *bottom-heat* in the rooting medium is also helpful to produce roots.

3. Use of Grafted and Budded Materials

Propagation by grafting or budding to raise trees and shrubs for planting in windbreak row is hardly a recommended practice, except in cases where there is a specific purpose to do so. This is because of the fact that grafted or budded trees usually become dwarf and compact in appearance and thereby, are less suitable for use as windbreak. These trees do not make extensively spread out root system and are also relatively less resistant to adversities on stress conditions. Further, weakening of the graft-joint even many years later, , i.e., or delayed symptoms of stionic incompatibility are met with in many cases. (Stion = Stock, *i.e.* rootstock+Scion)

4 Appropriate Method

What will be the appropriate method of propagation is an aspect that is largely dependent on plant species. Edapo-climatic situations and purpose of growing them are also important factors which need careful consideration. Kaith and Erasmus have recommended in their bulletin suitable propagation methods for some plant species used as windbreak, which is presented below.

(i) *Acacia arabica*: Direct sowing as seedlings do not withstand well rather than when transplanted.

(ii) *Agave americana*: Suckers or bulbils (to be planted on ridges).

(iii) *Anacardium occidentale*: Direct sowing.

(iv) *Azadirachta indica*: Transplanting of seedlings.

(v) *Cassia siamea*: Direct sowing or transplanting of seedlings.

(vi) *Dendrocalamus strictus*: Offset planting or transplanting of seedlings.

(vii) *Eucalyptus* species: Transplanting of seedlings.

(viii) *Gliricidia maculata*: Stem cuttings.

(ix) *Inga dulcis*: Direct sowing or transplanting of seedlings.

(x) *Jatropha curcas*: Stem cuttings

(xi) *Leucanea glauca*: Direct sowing.

9.8 Planting Season of Propagating Materials

It is generally advised that when seedlings or cuttings (= rooted cuttings) or layerlings (= rooted layers) are used as propagating materials to set up windbreak trees or shrubs, these should be planted at least a year earlier than when the fruit trees are planted in the orchard. The reason is, if planted earlier than the fruit trees, the windbreak trees will make higher growth than the former trees and accordingly, they will afford greater protection than when planted later. The viewpoint seeks justification in many circumstances but perhaps not in every situation. Sometimes difficulty may also be there to do so for practical reasons. However, establishment of the windbreak row(s) after planting of the fruit trees in the orchard is not a desirable practice and planting of the windbreak materials should at least be done simultaneously with planting of propagating materials for fruit trees in the orchard.

Regarding season, rainy-season is by far, the best time for sowing seeds, transplanting of seedlings or planting of other propagating materials in the windbreak row(s). Artificial irrigation may not be necessary at this season unless there appears a rainless period. In rainy-season, the plants also make greater growth but care should be taken to ensure that drainage is not impeded in the windbreak rows. However, apart from rainy-season, the propagating materials can also be set in spring-season when there is assured irrigation, or before that for many temperate zone plant materials.

9.9 Method of Planting

Where seedlings or cuttings or layerlings are used as propagating materials for windbreak trees or shrubs, the same method of planting as described under the Chapter 17 may be adopted. Broadly speaking, this consists of digging 90-100 cm cubic pits in summer and exposing the dug out soil to scorching heat in summer in order to kill insect larvae, disease organisms *etc.* Then the dug out soil should be thoroughly mixed with manure and returned this soil-manure mixture to the pit. After that, the filled-in pit should be watered every day to settle the mixture in it.

Application of the same method for planting of propagating materials of the windbreak materials is not, that is unwarrantable but so much sophistication is perhaps unnecessary for trees and shrubs used for setting in windbreak rows. From practical point of view, the positions where a row is to be made should be spaded upto about 15 cm depth and the soil overturned. After weathering, the turned out soil for some days, clods should be broken, soil should be thoroughly mixed with well-decomposed organic manure and leaf-mould, watered, flattened after drainage and then planted. Addition of *neem* cake in the row is definitely of advantage to get ride of white ants at the site.

9.10 Protecting Orchard Trees from Invasion of Windbreak Rows

Sometimes the windbreak trees make a knavish trick to cause deprivation of the fruit trees grown in the orchard. This is done by them in sending forth their roots inside the orchard soil. That is, when their roots make way to enter the soil in the orchard where the roots of the fruit trees abound, the water and the nutrients that are supplied to the fruit trees are actually pilfered by them through their feeder roots. As a result, the absorbing organs of a fruit tree, *i.e.*, its feeder roots are deprived of getting the inputs and accordingly, growth of it is hampered. The windbreak trees are usually wild in nature and accordingly, when they get moisture and nutrients in the soil, they withdraw voraciously. The roots of the fruit trees face competition with roots of the windbreak trees and in this competition, the latter usually wins. The orchardist cannot see this struggle among roots which happens underground and what is noticeable to him is improper growth of his fruit trees and vigorous growth of the windbreak trees, even though optimum supply of agro-inputs is ensured for the fruit trees in the orchard.

A simple device may however, be done to avert such encroachment of the roots of windbreak trees inside the orchard. As shown in Figure 9.7, this may be done by digging a long trench of about 1 metre depth between the windbreak row and the first row of the fruit trees (windward side). The trench should be dug at about 3 metres away from the last row of windbreak (leeward side).

The roots of windbreak trees that would tend to direct to the orchard soil would evidently become visible inside the trench. Now, the trench should be inspected regularly and whenever the roots of the windbreak trees would come to notice, these should be cut off with a chopping implement, or killed by applying kerosene or any other phytotoxic chemical. Later on, the trench may be filled up, if necessary and the same practice should be repeated after every two or three years.

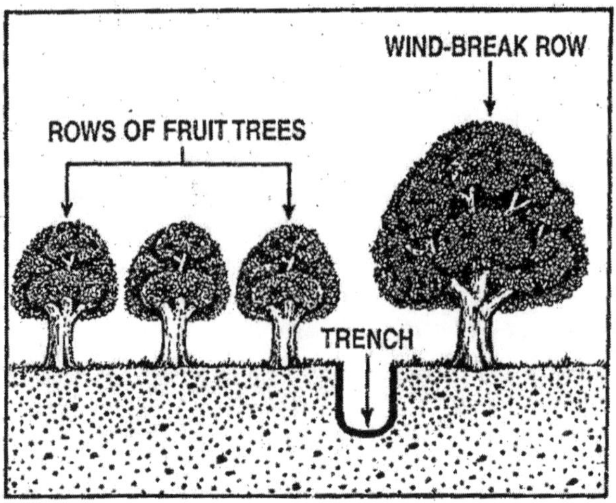

Figure 9.7: Trench Dug between the Windbreak Row and the First Row of Orchard Trees.

The above practice should be done in the secondary windbreak row also, if such a row has been set inside the orchard. In this case however, trench should be dug on both sides of the windbreak row since fruit trees exist on both sides of it.

9.11 Post-Planting Operations of Windbreak Trees and Shrubs

Unlike orchard trees and shrubs which need regular care as long as they are cropped, the windbreak trees and shrubs do not need so much care once they get established in their positions where they have been set. Nevertheless, to make them more effective, some operations are of great value. Among other operations, the following are considered of greater importance.

1. Training and Pruning

It should be remembered that not only for fruit trees but for windbreak trees also, *training* and in some cases *pruning* are necessary operations to be done.

Training in fact, refers to giving any *particular size, shape or form to a plant* and to do so, it may or may not become necessary to cut a plant part. By pruning is meant compulsorily cutting of any part of the plant, no matter it is a branch, twig, inflorescence, flower, fruit, leaf or root. In horticulture however, the term, pruning is applied when a plant part is cut off to affect *function* primarily, which most commonly relates to raise its *productivity*, either quantitatively, or qualitatively or both.

As stated earlier, for the purpose of planting in windbreak rows, such materials should be chosen which assume highly spreading, vigorous and tall appearance when fully grown-up. Thus, these trees would naturally become so in size and shape in their mature condition. But by suitable treatment, they could be brought to desired appearance in a more correct manner, which sometimes does not take place due to environmental conditions. Evidently, the properly performed training operation is the treatment which aims at so.

(a) Systems of Training

For erect growing trees and shrubs, three conventional systems of training are usually followed. These have been stated below:

(i) Central-leader,

(ii) Open-centre, *i.e. vase* and

(iii) Modified central-leader systems.

As shown in Figure 9.8 (a), in the central-leader system, the centre of the tree is allowed to lead and not the horizontally spreading branches. As a result, the tree assumes a tall appearance with less sidewise spread. The open-centre system is just the reverse, in which case, the height of the tree is less but spread is quite large [Figure 9.8(b)]. The modified central leader system [Figure 5.8(c)] is an intermediate form between the above two systems where the tree becomes neither very tall nor of very wide sidewise spread.

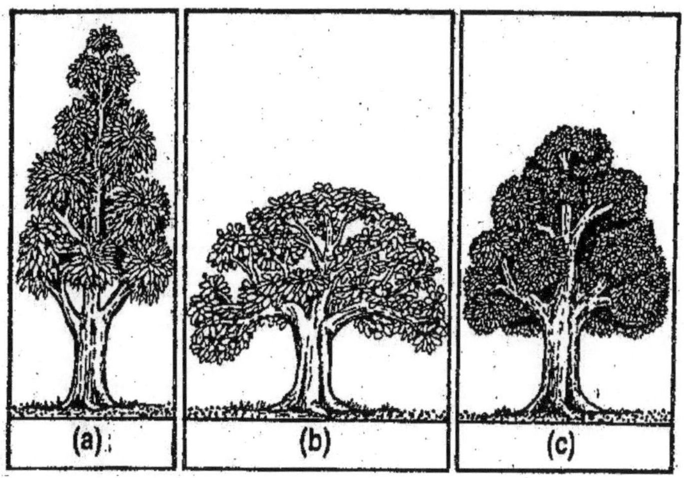

Figure 9.8: Systems of Training. (a) Central-leader system (b) Open-centre system and (c) Modified central leader system.

(b) Maturity of Trees for Training

It is to be kept in mind that whatever system of training that may be adopted, it should actually be started at *very early stage of growth* of the trees, that is, during the first 2-4 years of maturity depending on the species. This is because, in grown-up condition, it becomes extremely difficult or perhaps impossible to make any large change in the shape of a tree from what has been attained by it.

(c) Methods Involved

For the purpose to serve as a windbreak, a tree is most commonly trained by the open-centre or modified central-leader system and the central-leader system is usually considered unsuitable, except in special circumstances.

A tree is trained by the above-mentioned two systems by the regular procedure of *thinning-out* and *heading-back* techniques at its early stage of growth. Thinning-out operation refers to cutting a branch completely at its base in such a way that no part of it actually remains present there after it is cut [Figure 9.9(a)]. This operation is in fact, preformed where there is overcrowding of branches, *i.e.*, there is a congestion of small branches at one place. Among the closely spaced branches, some are selected and thinned-out to get rid of the congestion and this is done to ensure that the branches which are left uncut are well-spaced.

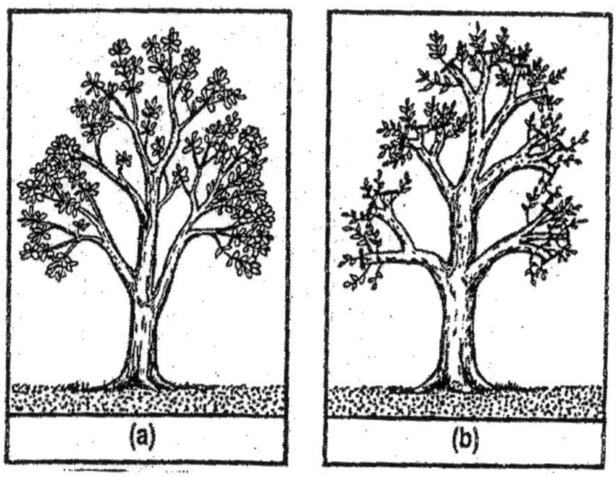

Figure 9.9: Trimming Techniques.
(a) Thinning out and (b) Heading-back.

On the other hand, the heading-back technique, which is reverse of it consists of cutting a branch only at its upper part and thus, the lower, *i.e.*, basal part of it is left intact [Figure 9.9(b)]. By cutting the upper part of the branch, the *apical dominance* of it is broken and thus, results in emergence and free growth of many side-shoots. Since many side-shoots come out with closeness at a place by this act of trimming, the technique is adopted at that part where branches are sparse, *i.e.*, less in number and thereby, this is done with the idea to force many new shoots.

(d) Training at Early Stage of Growth

The overall procedure to train a tree by the *open-centre* (= vase), or modified central-leader systems during the first few years of its growth has been outlined in the following.

In the first year, the tree should be left to grow freely in an undistrurbed manner, *i.e.*, no part of it is cut. At this stage, it should make good growth and to make then so, copious supply of agro-inputs like manure, fertilizers, irrigation etc. is essential. In the second year, if the growth is satisfactory, it should be de-topped at a suitable height (Figure 9.10).

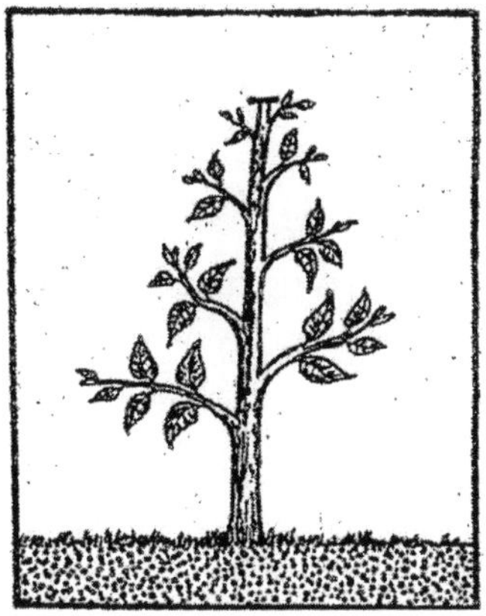

Figure 9.10: De-topping of Young Plant.

By the act of de-topping, many side-shoots would be produced from the trunk, both from its upper and lower parts and in different directions. These branches should be left to grow freely till they are 15-20 cm long. At this stage, 4-6 branches which have grown on all sides and are more or less equally spaced should only be selected to retain and the rest are cut off by thinning-out operation [Figure 9.9(a)]. Then these branches are headed-back [Figure 9.9(b)], leaving 10-15 cm of their basal parts intact.

From the headed-back branches, new branches would be seen to arise. Among these branches, 3-4 which have grown on all sides with suitable spacing should only be retained and the rest are cut back by thinning-out, while the rest are headed-back in the same way as before. New branches would again arise from the headed-back stubs and 2-3 of them radiated on all sides of a stub should be selected and headed-back while the rest are thinned-out. In this way, the operations are repeated during the early stage of growth of the trees, in order to build up their framework. The trimming of branches should better be done twice in a year, *i.e.*, before rainy-season and in winter.

It should be remembered that the branches which are upright, *i.e.* grow vertically upwards make the tree taller while the horizontally growing branches make the tree to assume a spreading form. Thus, to train the tree by open-centre (vase) system, horizontally growing branches should be greater in number than the upright branches, while in modified central-leader system, the latter branches should be a little more than the former. This should be taken into consideration in retaining or cutting of a branch when the branches are thinned-out or headed-back. In general, in open-centre system, not more than 20 per cent upright shoots are retained while in modified central-leader system, it may be 30 per cent but this would actually vary according to species. The central-leader system of training is usually uncalled for in that, in most cases, it is not desirable to build up the framework of a tree, if it is used as windbreak. Nevertheless, it is pointed out that under certain conditions, there may be necessity to train at least some windbreak trees by this system. To train a tree by central-leader system, the most important point is to encourage the main trunk to grow taller and taller and hence, no de-topping of it should be done. The branches should not be speready in nature, for which repeated heading-back should be avoided.

The general procedure for training trees for windbreak purpose during first few years of their growth has only been stated above and it should be remembered that the actual procedure that would be adopted depend much on many other factors. In fact, the branch which should be cut or retained should be judiciously selected by the orchardist himself.

(e) Training after Primary Stage

After the framework of the tree has been built up by suitable training procedure done during the first few years, much operations are not required during the subsequent years. However, some works may become necessary at this period and before the tree is grown-up. For example, it may be necessary to change a *crotch angle* to a desired manner. It is a common observation that if the crotch angle formed between two branches is narrow [Figure 9.11(a)] it becomes structurally weak and hence, the branches may break apart easily. Again, if the angle is very wide [Figure 9.11(b)] then also it becomes weak. Therefore, an angle of 50°-70° is considered optimum to provide optimum strength in most cases [Figure 9.11(c)].

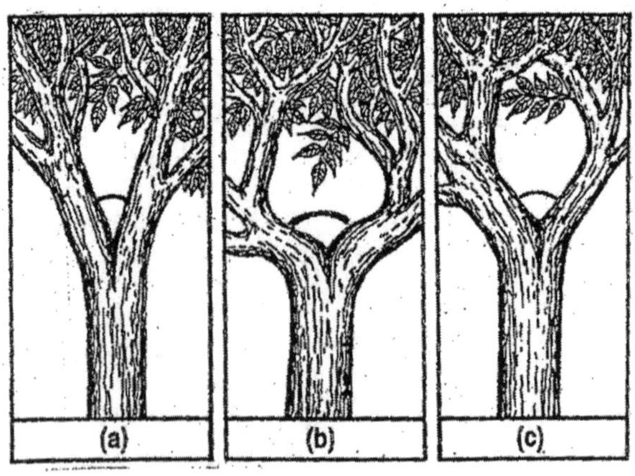

Figure 9.11: Angulation between Two Adjacent Branches. (a) Narrow, (b) Very wide and (c) Moderate (satisfactory).

Another important factor which gives strength to branches is relative weight of the two adjoining branches. If the two adjoining branches including their sub-branches and foliage parts are more or less of equal weight, strength is lesser and hence, the crotch between them may be more easily damaged or broken in high windstorm. On the other hand, if the weights of two adjoining branches are unequal, *greater strength is afforded at the junction* and the chance of their breakage due to wind force becomes less.

This is because of the fact that when there would be storm, the branch which is of lighter weight would bend towards the heavier one with less difficulty and hence, the stress that would be produced at the junction becomes much lesser. Therefore, it may be necessary to cut off a branch, if the adjoining branch forming the angle with it becomes smaller in thickness and weight. The operation should no doubt, be done with proper judgment.

(f) Removal of Flowers

It should be remembered that at the early stage of growth of a tree, the objective should be to encourage *more and more vegetative growth* in it. But this is sometimes hampered if flowers appear at this stage, *i.e.*, in premature condition of the tree. Production of flowers by a tree premature condition of the tree. Production of flowers by a tree prematurely may result especially when it is propagated by vegetative means. Disease condition may also sometimes result in prematureing flowering. Since production of flowers is antagonistic of a tree should be plucked immediately.

(g) Pruning

Pruning of a tree is in fact, done to effect its *function* and hence, it is an essential operation for temperate fruit crops as well as many subtropical and even some tropical fruits also to enhance their productivity. In that sense, the pruning operation appears to have little necessity for such trees which are planted as windbreak *i.e.*, with the idea to offer resistance against strong wind and not to get fruit from them.

However, as stated in the early part of this Chapter, some trees or shrubs or herbs (banana, plantain) which produce human consumable fruits are also sometimes planted as windbreak. In fact, such type of planting is done for more than one purpose, *i.e.*, silkworms (as in mulberries), lac-worms (as in ber), or bees, to secure leaves (as in saal, kendu), fuelwood, timber and the like. In such cases, pruning appears to be a necessary step. It may be noted that as the operation is concerned directly with production, the operation is done in the trees when they are in grown-up stage.

It should not be forgotten that in order to get the desired results, pruning should be done judiciously in consideration to season, amount of pruning and the parts to be pruned. For fruit trees and shrubs in particular, it should be kept in mind that there are many species which produce flowers and fruits in mature shoots while there are many which produce only in new, *i.e.*, current season's shoots. Thus, pruning should be done in such a way that such type of shoots which are productive should exist in abundance when flowering starts.

N.B. Whenever any branch is cut off, there is chance of getting it infected through the cut-surfaces. There is also the danger of drying or loss of adequate moisture from the cut-surfaces of stems, particularly when they are thick. Therefore, it is advisable to apply some *fungicidal paste* over the cut-surfaces of stem just after cutting. Besmearing the cut-surfaces with such paste is an effective means to prevent infection and also drying.

2. Application of Manures and Fertilizers

The trees and shrubs which are planted for windbreak purpose usually grow well in ordinary soil and without supply of manure or fertilizer. Nevertheless, they would definitely make vigorous growth, if the soil is made enriched by application of manure and fertilizers at least during the first few years of their growth.

During sowing or planting, preparation of the soil by incorporating well-rotten manure has been indicated in the earlier part of the Chapter. Besides this basal application, addition of manure, if done every year would undoubtedly render much benefit. It may not be essential to maintain any specific does but application of suitable amount of nitrogenous and phosphatic fertilizers along with lesser amount of potassium in the rainy-season would be satisfactory to boost up the growth of the windbreak trees or shrubs.

The amount of manure and fertilizers however, needs to be increased every year with progress of growth of the trees or shrubs till they attain 8-10 years of maturity. Application beyond this age if desired, should be done with the dose received in the 8th or 9th year.

Sometimes it is noticed that growth of the trees or shrubs is much less in some parts of the rows. In such situation, the manure and fertilizers are to be applied in higher amount in those spaces where growth looks lean.

3. Mulching

The act of covering the soil of the windbreak rows with mulch is undoubtedly of great help in conserving soil moisture and preventing weed growth. Organic manure, if left *in situ* would also act as manure beside mulching.

4. Irrigation

Liberal irrigations in summer and winter seasons during the first few years of growth of the windbreak trees and shrubs cannot also be disregarded to encourage growth of them. If manure and fertilizers are applied at the rows, irrigation needs to be given after application.

5. Plant Protection Measures

Plant protection measures are among other supplement that would benefit, not only the windbreak trees and shrubs but also the fruit trees grown in the orchard.

The windbreak trees and shrubs sometimes harbour many insect pests and diseases and these also attack the fruit trees located in the orchard. Hence, it is necessary to control them in order to save the fruit trees. It should also be remembered that sometimes the windbreak trees become the roosting places of birds which damage the fruits in the orchard and arrangement to scare them should be made.

Chapter 10

Orchard Accessories

10.1 Importance

Apart from growing fruit crops and inter-crop(s) if any, sometimes it may become necessary to provide even some non-crop items also in the orchard. These items, which may be designated as *accessories* are included in the orchard with the purpose that cropping and management of the same could rather be done in a more efficient manner. Thus, although these accessories are not directly associated with cropping, they have *auxiliary function* in running the orchard in the desired direction. That is, they aid in running the orchard in more successful manner. If the orchard is a small one, it may not become so necessary to include the different accessories in it but for a fairly large-sized orchard where activities are numerous, inclusion of some of the accessories cannot be disregarded. Nevertheless, it can be said that at least a few accessories appear to have much necessity even if the orchard is a small-sized one.

In fact, there may be quite a number of accessories which have justification for inclusion in the orchard. But from the practical point of view, which accessory would be fitting for which orchard, is a matter of judgement to be decided by the would-be orchardist. He should especially take into consideration some aspects notably, location of the orchard, its surroundings, whether other orchards exist around it, attitude of the people in the locality, security, area of the orchard, its soil and climatic conditions, crops and inter-crops to be grown, method of culture to be adopted, management system to be followed, labourers and other staff to be employed, manner of disposal of the produce to be done, market facility, funding capacity of the orchrdist and so on.

It should be kept in mind that provision of an accessory would evidently necessitate some extra expenditure besides the fact that, some valuable land in the orchard may also become unproductive by its installation, giving no return. Therefore, an accessory should be included in the orchard only when it has any real purpose and not just out of fancy, particularly when it has been decided to run the orchard on commercial basis, where economy is the primary consideration.

10.2 Particulars of Accessories

Quite a large number of accessories are in fact known, each of which has some advantage when provided in an orchard. The list will be very long if all of them are brought to notice and discussed. Therefore, only those accessories that are usually of much greater use have been enlisted below stating their merits and proper locations in the orchard where provision of those should be made. It is to be noted that while some of the accessories have necessity in orchard of any size, other items find necessity when the orchard is large or very large in size. Some of the accessories may be required temporarily or seasonally or at certain phase of growth of the orchard. It is suggested that the orchardist should provide an accessory in his orchard only when he considers genuine need of that. In some cases, he should seek advice of other persons also in this respect, who are expert in orcharding and have practical experience.

1. Bed of Lime Dust

Provision of lime dust close to each gate within the orchard appears to be a good step. The idea of this practice lies in the fact that whoever enters the orchard through any of the gate cannot do so without walking over the layer of lime dust. The lime dust should be applied on soil surface covering a lengthwise distance of 3-4 metres. In order that the lime particles remain present in its place and is not washed away with water, it is better to make a small wall of 20-25 cm height and pour the lime inside it.

The idea of the above practice lies in that, the orchard might be protected against entry of insect eggs, termites, disease organisms *etc.* inside it through the shoes of the persons and at least to some extent. It should be remembered that not only insects or animals contaminate pests or diseases but human beings are also no exception. (There is a story. In a tobacco farm, a person, say Mr. X was working in the farm when he met a friend outside. He rushed outside his tobacco farm and shook hands with his friend, say Mr. Y. After sometime, Mr. Y left the place and he met another friend of him, say Mr. Z on the way and he shook hands with Mr. Z. Mr. Z was a tobacco farmer. He went to his farm and began to prune his tobacco plants. It was found that his healthy tobacco plant were infected with the virus disease which is known as *Tobacco Mosaic Virus, i.e.,* TMV disease. On investigation, it was learnt that the tobacco plants of the farm of Mr. X were already contaminated with the said virus and those virus particles actually made contamination of the tobacco plants of the farm of Mr. Z by hand).

2 Irrigation Arrangement

Artificial irrigation is indispensably required for scientific fruit culture in most parts of India and accordingly, this necessitates arrangement for ensured supply of irrigation water in the orchard whenever and to whatever amount it is required. The source from which irrigation water is obtained may be located either outside the orchard or inside of it.

If the source of water lies outside the orchard, arrangement to convey the water from the point of inlet and its delivery upto the cropping plots may be all that is necessary. The water may be delivered to the plots through earthen channels, pipe-lines *etc.* in the orchard. Masonary type of irrigation channel (Figure 10.1) which was more prevalent in olden days may also be constructed with one or two layers of bricks. The channel should have a width of about 30 cm and a depth of 40 cm and it should remain at a higher level than the ground, such that the irrigation water could be easily let out to the plots by gravitational flow.

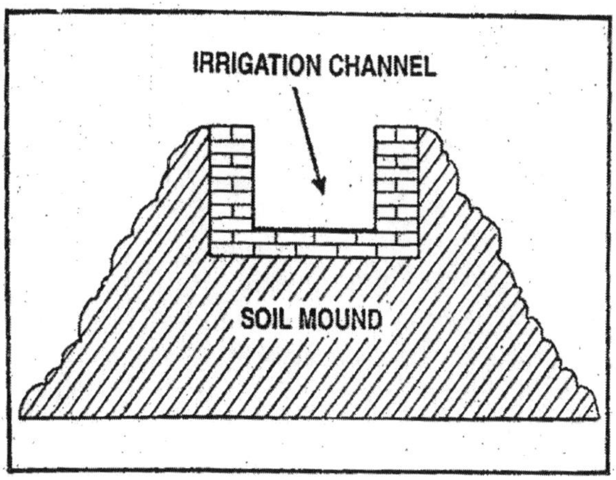

Figure 10.1: Irrigation Channel of Masonary Type.

On the other hand, if the irrigation water is not obtainable from outside source, arrangement should evidently be made for its availability inside the orchard. For this purpose, water source needs to be set up within the orchard area by constructing deep or shallow tube-wells, or by digging wells, tanks *etc.* as the case may be.

A deep tube-well which can irrigate a large area of land may be set up in the central part of the orchard. The shallow tube-well, which if required more than one in number should better be located in a scattered manner at different sectors in the orchard, instead of concentrating those at one place.

In the recent times, *drip* and *sprinkler* systems of irrigation have gained a breakthrough, especially in the irrigation of fruit crops. If these systems of irrigation are planned to be installed in the orchard, special arrangements are also to be made for their setting up. These would require installation of suitable pumping arrangement to impel flow of water or provision of overhead tank may also become necessary to operate these systems. (Ref.: Mazumdar, B.C. 2022. *Orchard Irrigation and Soil Management Practices.* Daya Publishing House, New Delhi-110002)

3. Roads, Paths and Alleys

For easy, quick and unobstructed transport of machineries, heavy implements or bulk of manures, fertilizers *etc.*, to a distant cropping plot in a large-sized orchard and bringing back the harvested produce from the cropping plots to the godown or elsewhere, means of transport with wheels are essential in the from of hand or bullock-driven carts, pedi-vans, motor vans, tractors with trailors or trucks, small or large *etc.* Obviously, this necessitates provision of sitable roads, paths *etc.*, in the orchard for their easy movement.

The network of roads, paths, alleys *etc.* that are to be laid out in an orchard would no doubt, depend on a number of factors. However, to make economy of land, their layout to reach each and every cropping plot in the orchard may be dispensed with. In fact, the purpose is served well if the different sectors in the orchard are approachable the roads or paths.

Making all the roads of same width is also unnecessary and this would rather make wastage of much land in the orchard. Similarly, there is no essentiality for all the roads to get metalled. In fact, unmetalled paths or alleys may be laid out where heavy vehicles are not needed to be moved.

4. Manure Pits

Many orchardists are not in favour of constructing any manure pit in the orchard for according to them, these are liable to affect sanitary condition of the orchard. However, to dispose of fallen leaves and pruned parts of the fruit trees, unused parts of the inter-crops, damaged fruits *etc.* easily and quickly and also to get some compost ready-at-hand, necessity of manure pits in the orchard cannot be disregarded. Pits for storing dung of bullocks should also be made. These apart, it is also advisable to make vermicompost in the orchard. It is important that manure pits should not be located close to the cropping plots in the orchard.

5. Office

When the orchard is quite large, activities are also numerous and the necessity of a room for official works is keenly felt in such case. The room may not be a large one and should be so spacious as can accommodate a working table, few chairs and one or two racks, almirahs to maintain files, registers, ledgers, cash-book, stock-book *etc.* A fire-extinguishing equipment should also be maintained in the office at such a place that is visible and within easy reach.

The central part of the orchard and the road side appears to be a suitable site for the office room. Many orchrdists however, prefer to locate it near the main entrance of the orchard.

6. Storehouse for Agrochemicals

In a large-sized orchard, it is advisable that fertilizers which are needed to be used should be stored in bulk, instead of purchasing them from the market at the time of use. For this purpose, it is necessary to maintain a separate godown in the orchard. Since fertilizers are needed to be stored in dry place, the godown should be located in such a situation where there is no dampness. In construction of the godown also, arrangement should be made that the walls and floor of it do not become damp in rainy season. For convenience in storing different kinds of fertilizers, suitable shelves should be provided in the godown.

Apart from fertilizers, other agrochemicals also become necessary to be stored for use in the orchard. For this purpose, it is advisable to maintain two small annexes at both sides of the fertilizer godown. One of these annexes should be used for storing crop protection chemicals while the other for micro-nutrient salts, plant growth regulators *etc.* Apparatuses required to make solutions of the above materials should also be stored in annexes or in the laboratory.

7. Shed for Manures and Soil-Amendments

It is advisable not to store farmyard manure of compost in the orchard as these are likely to harbour many pests. However, oil-cakes, bonemeal *etc.*, are safer in this respect and may be stored for some period. Accordingly, a shed similar to that of machineries is needed to be constructed with compartments for storing the materials separately.

In the same shed, separate compartments may also be made for storing soil-amendments, *e.g.*, lime, gypsum, sulphur *etc.*, if these are required to be used in the orchard.

The suitable sites for storehouse of chemicals, machineries and manures may be at some distance away from the office-building with one after the other.

8. Shed for Machineries and Vehicles

In order to maintain these materials, a separate shed is necessary to be maintained in the orchard, the roof of which may be of tin or plastic (asbestos should not be used as it is carcinogemic) sheets. Handling of these equipments would become convenient if the shed is divided into compartments such that materials of one kind be maintained in a separate compartment. Thus, in accordance with the materials that are maintained, separate compartments may be made, say for truck, tractor, motor van, trailor, cart, van, power-sprayer and duster, pumping set as well as heavy and light implements.

9. Shed for Bullocks

If bullocks are maintained for ploughing or other soil operations, suitable sheds should be maintained for them. Normally, a floor space of 3.5 metre long and 2 metre wide becomes necessary for one bullock and in addition, space for forage and water are also necessary. Bullock shed should be allocated near the compost manure pit or farmyard manure pit. (Bullocks may not be required, if tractors, small tractors or power-tilleres are used).

10. Storehouse for Harvested Produce

The produce harvested from the fruit crops or other inter-crops grown in the orchard may be necessary to be stored for some period. Accordingly, separate storehouse(s) needs to be maintained in the orchard for storing these produce. The storehouse for this purpose should however, be specially constructed taking into consideration the following aspects.

The most important point is, the storehouse should not have any dampness. Therefore, it should not be built up in a damp place and also that the walls or floor of this house should be of such material as would ensure it to be damp-proof in rainy-season.

It is ideal if refrigeration is maintained where fruits are stored, failing which, care should be taken that the storehouse does not become very hot. For this, tin or plastic roofing is not desirable. If these materials are at all used, these should be covered with insulating materials. In building up the walls, use of hollow-bricks is also known to prevent the room in getting very hot in summer season.

Washing of the store becomes necessary occasionally. Accordingly, its walls and floor should be coated with water-proofing material which would permit washing those with soap water, disinfecting liquid *etc.* It is better if the floor is made somewhat slopy so that water drains out easily and quickly after washing.

Ventilation of the storehouse is an important aspect in storing fruits and to ensure this, the store should be provided with windows, fans *etc.* It is pointed out in this connection that many fruits are known to emit ethylene gas under post-harvest condition. This hydrocarbon gas, going by the term, *ripening hormone* is given off by such fruits especially during or just before ripening which mayu be and causes the unripe fruits to ripen quickly on exposure to even very small amount. This is not desirable as early ripening lowers the shelf-life of fruits. Therefore, to prevent ripening induced by ethylene exposure, suitable arrangement should be made to drive out this gas that accumulates inside the storehouse, emanating from fruits. Occasional fanning out the gas serves this purpose to some extent.

Another important aspect that should not escape attention is to make suitable device to prevent entry of rats, squirrels *etc.* into the godown. A protective device may be to construct a suitable drain around the godown on all sides. Water mixed with toxic or pest repellent chemicals may be poured into the drain due to which the walking animals cannot cross the drain and enter in the godown (Figure 10.2).

Some orchardists also prefer to construct a basement with the idea that some of the produce are stored at the above-ground godown while those which are better stored in cool condition are maintained at the basement which has been costructed under the above-ground godown.

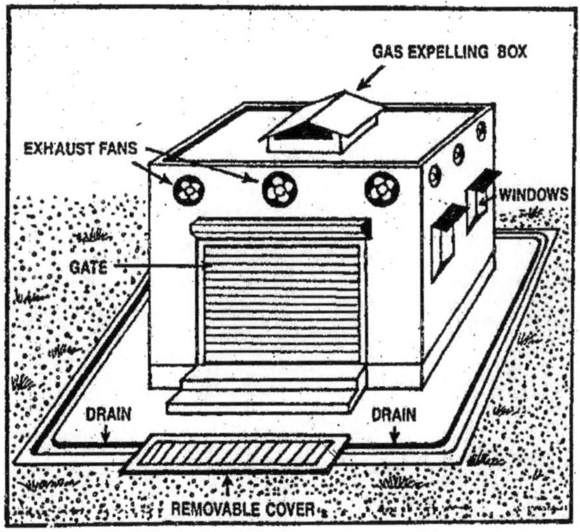

Figure 10.2: Storehouse for Harvested Produce.

In regard to site, the storehouse for harvested produce should be located away from the sheds where manure, fertilizers, implements are stored. The other side of the office building could however, be chosen as a suitable site for this purpose.

11. Drying and Threshing Floor

Some fruits and nuts are there for which, sun-drying for some period after harvest becomes necessary. Threshing may also become necessary for some inter-crops, if grown in the orchard. For such post-harvest operations, a concrete floor raised to few centimetres above ground level becomes necessary to be constructed in the orchard.

The space at the back side of the office building may be a suitable site to construct this flat floor.

It is desirable if the concrete floor is made somewhat large in size, such that part of half of it remains open while the remaining part is provided with roof of tin sheets or even tarpaulin (Figure 10.3). The purpose of making this lies in the fact that in case of sudden rain or windstorm, the materials left for drying or threshing on the open part of the floor may be quickly brought to the extension part of the floor that has roof over it.

The roofed part of the floor may also be used as a conference hall for the orchard workers when needed and construction of a raised platform at the end of this part would serve the purpose in a better way. This part may also be used by the workers for recreational activities like watching television or video programmes, opera shows, dramatic performances, indoor games *etc.*

It is also desirable to construct one or two rooms in the roofed part, which may be at the back side of the platform, where ripening may be done for those fruits which require post-harvest ripening, *e.g.*, banana, mango, sapota, jackfruit, peach *etc.*

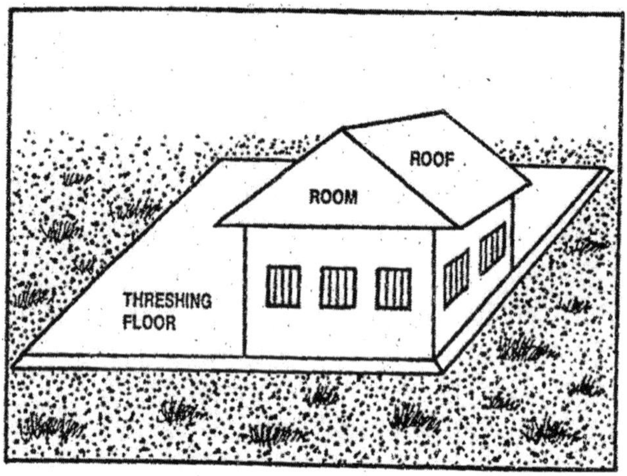

Figure 10.3: Drying and Threshing Floor.

12. Nursery

Maintaining a nursery in the orchard is very much helpful for carrying out some propagation works which are needed to be performed quickly and are best done within the orchard itself. A large or a fully-fledged nursery though not required, a small-scale and working type may be all that is necessary.

A partial shady situation in the orchard would be a suitable site for the nursery and the area near the windbreak may also be considered for the purpose.

13. Workers' Quarters

Where the orchard is a considerably large one, plentiful activities are needed to be performed throughout the year. For this, it is desirable that at least some of the employees such as Orchard Manager, Assistant Manager(s), Field Overseer(s), Mate(s) and permanent laboures, whose services may be required at any time or at different hours of the day should reside within the orchard compound. Accordingly, residential accommodation should be provided to them, necessitating construction of quarters within the compound. About 140 square metre area may become necessary to construct a quarter of one labourer.

Many orchard owners are in favour of constructing residential quarters for all the staff in the central part of the orchard while many others consider it proper to build up the quarters of the Orchard Manager and the senior supervising staff in the central part and for others, near the boundary or at the corner places of the orchard. It is also held by many to build up all the quarters in a scattered manner all over the orchard with the idea that all parts could be under watch.

14. Resting Shed for Workers

Provision of resting places where field workers can take shelter during rain or scorching heat of midday or at their leisure have their necessity in the orchard. For this, it is necessary to construct sheds at different sectors in it. The sheds should also have amenities like drinking water, toilets, benches, cots *etc.*

15. First-aid Room

There is justification to make provision of a room for first-aid treatment of the staff when it becomes necessary. The room should preferably be made adjacent to the office. It should contain a table, 2-3 chairs, a bed, emergency medicines, bandage and other appliances that may be considered necessary to maintain.

16. Bathroom and Lavatory

Provision of toilet, washing arrangement, lavatory *etc.* in the orchard separately for male and female workers cannot be ignored.

17. Watch-Towers

Where the orchard is of very large size and where there is a chronic danger of virulent and large-scale theft of the orchard produce or the properties, the orchardist should evidently strengthen the secutiry system. The number of guards need to be increased in such case and they should keep vigil from high above the ground, especially at night. For this purpose, suitable watch-towers should be ereced at several points along the boundaries of the orchard. Watch towers may be of different types (Figure 10.4).

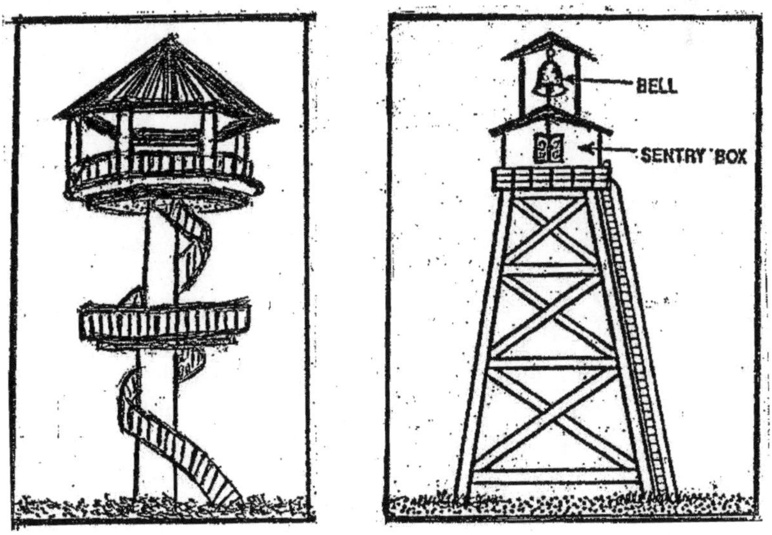

Figure 10.4: Watch-Towers.

18. Sentry-box

For security and prevention of entry of unwanted persons as also animals, a sentry-box for the guard at the entrance of the orchard has its essentiality. If there is more than one entrance, sentry-box should evidently be built up at each entrance.

19. Bird Scaring Device

Among animal pests, birds are known to cause a great loss not only by eating the fruits but also damaging a large number. Besides, the woodpeckers cause serious damage to the trees by clinging to the bark and pecking those to find insects and building nests.

As protective measures against birds, scarecrows may be hung on bamboo poles at several places in the orchard. Hanging bells on bamboo poles in a scattered manner all over the orchard and ringing those occasionally by pulling rope from ground, the other ends of which are tied to the bells is also an effective method. The cheap type of bells may however, be prepared by using old and damaged kerosene tins. The tins may be hung with their open ends facing downwards and for ringing, suitable objects like iron rods may be attached in the inside of a tin with rope passing through the hole made at the other side of the tin (Hanging black polythene bags at several places is also partly effective).

20. Weigh-bridge

Provision of one or more weigh-bridges have their necessity in the orchard for weighing heavy articles, loaded vehicles *etc.*

21. Siren

In order to signal the duty hours of the orchard workers in a very large orchard, it is expedient to install a siren system rather than bell(s), the ringing of which is difficult to be laudable at distant places.

22. Closed-Circuit Television Camera

For emergency services and rescue workers of staff and security purpose, installation of video surveillance system may be done in the orchard, particularly when the orchard area is large.

23. Microphone

Installation of microphone in the orchard office with loud-speakers posted at different places in the orchard for the purpose of communicating message to the workers who work at distant places has definitely been deserving and calls attention when the orchard occupies an immense area of land.

24. Guide-posts

It would be very much convenient for orchard employees, transport personnels and others, if suitably displayed guide-posts are featured prominently at different parts of the orchard.

(25) Visitors' Room

In a well-managed, large-sized orchard where many types of fruit crops are grown, people in the locality and especially the neighbouring orchardists feel interested to visit it occasionally and discuss with the orchard personnels to gather knowledge on matters relating to the crops and their management. As exchange of ideas always benefits both sides, the orchardist should welcome others in his orchard and for this purpose, he should consider of maintaining a room where visitors can comfortably sit together and have discussion. Location of this room adjacent to the office-room appears to be satisfactory.

26. Conference shed

If the orchard is very large, a separate place with plastic roof may be allocated to hold conferences with neighbouring orchardists, experts and government officers *etc.* However, this does not appear to be very necessary to make provision of such a separate place as the drying and threshing floor may even be used for the purpose as has been mentioned under the Heading, 10.

27. Metrological Observatory

Maintaining even a small type of meteorological observatory in the orchard has its justification. If sophisticated equipments are not possible to be installed, a few items, such as a rain-gauge, a thermometer to record maximum and minimum temperatures, a barometer, an anemometer, an illuminator *etc.* may be maintained in it.

28. Electric and Telephone Lines

The electrical and telephonic facilities in the orchard need no emphasis. For this, it may be necessary to erect posts within the orchard. To provide lighting facility at different parts of the orchard, electrical posts should also be erected accordingly. Some important telephone numbers like those of fire stations police-station, ambulance centres, physicians *etc.* should be displayed clearly not only in the office but also wherever telephone-booths have been installed in the orchard.

29. Library

For carrying out orchard operations in scientific manner, sometimes it becomes necessary for the orchard personnels to consult books, magazines *etc.*, to have at hand informations. For this purpose, maintaining even a small-sized library with reading table has its usefulness in the orchard.

Large number of collections though may not become necessary, some standard books on the culture of fruit and other crops that are grown in the orchard are enough for maintenance. Inclusion of some magazines, bulletins, periodicals or even journals which provide informations on the recently developed technology on the culture of fruit and other crops needs emphasis. A room adjacent to the office-room may be a suitable site for the library room.

30. Laboratory

In a modern large-sized orchard of international standard that runs on scientific principles, maintaining a laboratory seeks Justification in it. The laboratory may not be a large one and even a small working type may be enough.

The idea to set up such a laboratory would be to make analyses of some components of fruits (Ref.: Mazumder B.C. and Mazumder K. 2003. *Methods on Physico Chemical Analysis of Fruits.* Daya Publishing House, New Delhi-110002), leaves, petioles *etc.* and also soil samples. The orchard operations could no doubt, be carried out on scientific basis by applying the techniques based on results after performing the analyses.

Existence of such a laboratory is certainly useful as the analysis of samples could be done personally by the orchard personnels and directly in the orchard, without sending to other place or depending on others. It is also pointed out in this connection that instead of setting up a laboratory separately in each and every orchard in the locality, maintaining one in an orchard may even cater to the need of many surrounding orchards and if required, on payment basis also.

Without making much sophistication, only some selected equipments like a physical and a chemical balance, a colorimeter, a centrifugal machine, a small dryer, a hand-refractometer, a hand *p*H meter, micro-Kjeldahl distillation set, a simple microscope and necessary glass or plastic goods, chemicals, reagents, dyes, a distilled water set *etc.* are enough that may be maintained. The suitable site for the laboratory should be adjacent to the office-room.

It would perhaps be convenient if the office-room, visitors' room, library and the laboratory are all housed in the same building with partition walls between one another and making provision of one or two toilets also in the same building. If any car or jeep or a tempo-van, cycle-cart *etc.* is maintained, the garage for the same may also be located in this unit.

31. Sales Counter

There are many orchardists who consider it justifiable to sell some of the produce directly to its employees and even at subsidized price. Selling at least part of the produce to the outsiders directly from the orchard is also a choice of many orchardists in commercial orchards, particularly in consideration to the social aspect. For such type of selling in large-scale if desired, there is necessity to open up a sales counter in the orchard. Scrap or unused implements or other materials of the orchard, if desired to sell or auction may also be done from this counter. Location of the counter may be near the orchard entrance.

32. Canteen and store

Especially in the absence of such facilities at nearby place outside the orchard, the orchardist may consider in running a canteen for refreshment of the workers and also a general store to sell grocery, stationery articles *etc.* to them.

33. Beekeeping Arrangement

Apiculture in the orchard is undoubtedly of great benefit. It does not occupy space but gives honey and the orchardist earns extra money. More importantly, the practice is very much helpful to ensure pollination of entomophilous flowers which belong to many species of fruit trees (The bees are not crop pests as they steal only the nectar from flowers which does not hamper fruiting but man is a pest to them and steals their hard-earned reserve which he cannot prepare by himself).

34. Flower Gardens

A well-kept decorative garden in a working place is a likeable feature to everybody. In a spacious orchard also, one or more flower gardens are charming sites. In a commercial orchard however, flower gardens should not compete with the cropped area. If designed properly, a pleasing garden can be laid out even in a small area of land.

The place opposite to the office building is a suitable site to layout the garden, or the office building itself may be surrounded by the garden. A lawn, flower beds and even lily pool, fountain, statue *etc.*, may be added as the features of the garden. Road junctions and other available spaces may also be used to layout ornamental gardens.

Chapter 11

Overall Layout

11.1 Importance

After it is decided which fruit crops and inter-crops are to be grown in the orchard and which accessories are actually to be provided in it, the next step for the would-be orchardist should be to make an *overall layout* of the proposed orchard, which are broadly of the following two categories.

(i) Cropping plots.

(ii) Accessories.

Beyond any doubt, this is an essential step in setting up of an orchard and should be performed with a great extent of reasoning and rationality. This is because, the orchard is established on a long-term basis and once the cropping plots or the accessories have been set up in it, it is very difficult to change their positions in future. If at all, any large change is attempted to be accomplished during running condition of the orchard, it involves high expenditure and hampers its productiveness, besides disturbing the outfits of it. To avoid such hazards, allocation of the components should be done in a very intelligent manner and with assertiveness that it would not become necessary to make drastic change of their positions in future except in very special circumstances.

11.2 Factors Affecting Layout

If the orchard is run on experimental, teaching or such other academic purposes or to meet the requirement of a nursery, an organization *etc.*, profit may not be the primary consideration. But if it is run on a commercial basis, securing profit is by far, the sole objective and this is accomplished by heightening the productiveness, either quantitatively or qualitatively, or both by application of appropriate agro-inputs and efficient management. With heedful consideration to this basic principle, the overall architecture of the orchard should be planned. Beyond doubt, this layout is a very important part in achieving the objective of the orchard, necessitating consideration of a number of factors, the important of which have been laid down in the following.

1. Preference of Flat Land

To set up and orchard, a land which is flat is considered best. In undulated or slopy land, culture of fruit crops is difficult to be done with success. Under such edaphic condition, the moisture and plant nutrients present in solution form in the soil will tend to move from the elevated to the depressed part. As a result, fruit crops planted on the elevated part will be deprived. But more harmful effect will reach to those fruit crops that have been planted in the depressed part of the land. This is because, unless the soil is of porous texture, accumulation of higher amount of moisture in that part eventuating poor drainage will seriously jeopardise the growth and productivity of the fruit crops there. In fruit culture, there is in fact, a generality which states, *even though other adversities in the soil are possible to withstand by the fruit crops but never poor drainage.*

Where the land is uneven, that should be flattened by shifting of soil in a careful manner. However, where slope is very sharp, it may not be practicable to flatten it by soil shifting process. In such type of land, terraces are advisable by making suitable bunds. As a matter of fact, in hilly area, terracing is almost an essential operation if orcharding is to be done. Where and how terraces should be done is no doubt, an aspect that needs great skill and layout of orchard should be made accordingly.

2. Size of Plot

Size of the cropping plot is another factor of great importance. It may be said that in general, small-sized cropping plots do not have much advantage in commercial orcharding and results in wastage of land.

3. Separate Plot in Mixed-Fruit Cropping

If mixed-fruit cropping, *i.e.*, culture of more than one species of fruit crops is desired in the same orchard as described later, separate plot should be made for each species, instead of jumbling together trees of different species in the same cropping plot.

4. Fruit Plants of different Heights

If more than one species of fruit crops which attain different heights on maturity are grown in the orchard, it will be justifiable to plant those in such a manner that the dwarf trees are set at the eastern side and the taller ones sequentially at the western side and thus, a gradient is maintained. As shown in Figure 11.1, by adopting this system of planting, all the trees will get adequate sunshine and the dwarf trees will not be shaded by the taller ones with movement of sun from the east to the west. In the afternoon hours, although the former trees may be shaded by the latter but not in the morning and the noon, the ray at which time has greater significance.

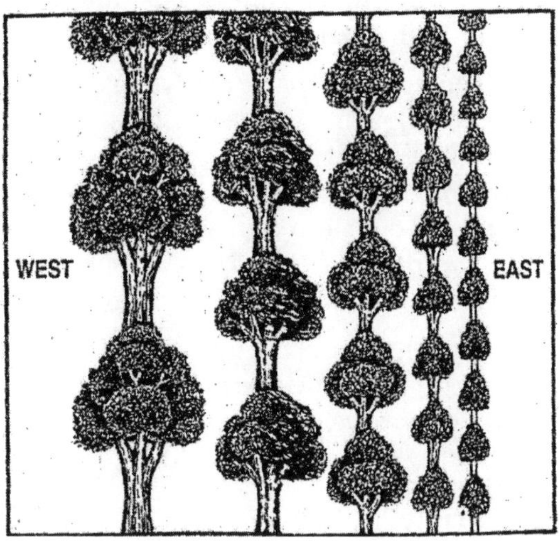

Figure 11.1: Planting of Fruit Trees with Dwarf Ones at the Eastern and Taller Ones at the Western Side, Maintaining Gradient.

5. Fruiting Time

In allocation of cropping plots when more than one species of fruit crops are grown, another aspect to be taken into consideration that is, those having same fruiting time should be set in plots which have nearness to one another. By this allocation, cultural operations of the trees, harvesting of fruits *etc.* become convenient, besides the fact that the loss of fruits by birds is not heavy for those belonging to one species.

6. Self-unfruitfulness

For some fuit crops, growing of more than one variety becomes essential for their fruiting. This is especially true for most of the temperate zone fruits. In such fruit trees, some varieties are self-unfruitful, which means that they set fruits when their flowers receive pollens from the flowers of trees that belonged to other varieties. For example, in apple the triploid varieties viz., Cox's Orange Pippin, Beauty of Bath, Baldwin, Bisbel Giant Winsap, Charlotte, Blenheim Orange, etc., are known to be self-unfruitful and hence, they require self- fruitful, diploid varieties like Golden Delicious, Red Delicious, Gloster, McIntosh, Jonathan, Rome Beauty, Allington Pippin, Newton Wonder, etc., as pollenizers i.e., to supply pollens to their flowers. In pears, most of the varieties are self-unfruitful although plenty of pollens are -produced by their flowers and fruits are set in them when cross-pollinated by other varieties. The variety, J.H. Hale in peach is also self-unfruitful and needs pollens from other varieties. Many cultivars of plum, cherry, black currant, chestnut, filberts have this character as well. It further needs mentioning in this connection that not only in temperate fruits but in some tropical and sub-tropicalfruits also, this character is noticeable. The famous mango variety, Dusheri may be cited as an example which is noted to be self- unfruitful under north Indian conditions and requires cross- pollination with other varieties.

In many varieties of fruit trees and particularly those of temperate zone group, it is also observed that even though they are self-fruitful, percentage of fruit set, retention of the fruits without premature fall as well as size of the fruits increase when cross-pollinated by other varieties. Thus, it is to be kept in mind that under the above conditions, not only one but more than one variety is necessary to be grown in the orchard. Evidently, those *pollenizing varieties* should be planted in plots close to the main varieties, in order that supply of pollens to the latter becomes easier.

With heedful consideration to the above-mentioned facts, layout of the orchard should be done such that cropping plots are provided for the pollenizer varieties and that these plots are close or even adjacent to the cropping plots of the main varieties.

7. More than One Variety

Growing more than one variety in the same plot of land is usually not desirable. This is particularly unwarrantable, if vigour and spread of trees of different varieties are not same when they are grown up. This is because, if trees of one variety are highly vigorous and occupy more space, those will suppress the trees of the other variety in having proper growth and assuming normal healthy appearance.

However, in some cases, mixture of varieties in the same plot of land is recommendable. For example, in growing temperate fruit trees which are self-unfruitful, as stated before, maintaining trees of any pollenizer variety becomes necessary without which, the flowers of the former will not set fruits. Many fruit scientists therefore, advise that the trees of the pollenizer variety should be set in the same plot of land where the trees of the self-unfruitful variety are set and the trees of the latter variety should in fact, be interplanted with trees of the former variety, with the idea that quick and timely supply of pollens be ensured. This may not become possible, if the pollenizer variety is grown in another plot, whatever close it may be to that of the plot where the self-unfruitful main variety is grown.

The above discussion would thus, make it apparent that a number of factors need to be considered in allocating cropping plots in the orchard. Some of these have only been presented above and these apart, there are also many other considerations that would depend in respect of the concerned orchard.

The allocation of orchard accessories is the other part in laying out an orchard having equal importance and has been discussed earlier (Chapter 10).

Judicious allocation of the cropping plots on the one hand and accessories on the other and their concordance give rise to overall layout of the orchard in scientific perspective.

11.3 Sketch of Layout

With due consideration to the different factors, the proposed orchardist should sit with pencil, eraser and paper to make a rough sketch of the layout. The survey of the orchard an land had already been done and using a copy of that map, the rough sketch should be made. Evidently, the sketch should be made to the *scale* so that actual measurements are obtained. Repeated drawing and erasing may become necessary to bring the drawing ready for the final shape.

However, before drawing the final sketch, it is also advisable that the proposed orchardist should take counsel with technical experts and other orchardists for any suggestion. Lastly, when he is fully satisfied, he should start for drawing the final map, which should be done to the scale and no doubt, by an efficient draftsman. Although some alterations of the plan may be done while setting up the orchard, any large change in it should be avoided.

11.4 Types of Layout

There is no single type of overall layout of an orchard that can be done. In fact, quite a number of layout is possible for a given land meant for orcharding. The sketch given below (Figure 11.2) is a common one made for a flat land which is square or rectangular in shape.

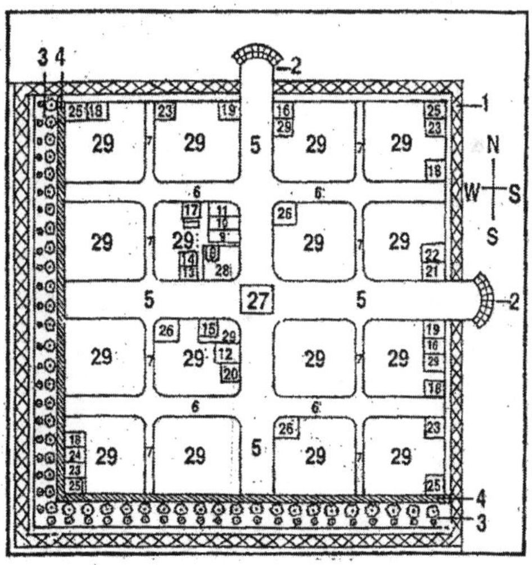

Figure 11.2: Plan of Overall Layout of a Large-Sized Orchard.

(*Index: 1:* Hedge, fence or other barrier, singly or combined; 2: Entrance, exit; 3: Wind-break rows; 4: Tench, 5: Main road; 6: Sub-road; 7: Alley, 8: Office, visitors'room, library, laboratory, toilets; 9: Chemical store; 10: Machinery and vehicle shed; 11: Manure shed: 12: Storehouse, 13: Drying and threshing floor; 14: Ripening room; 15: Orchard Manager's quarter; 16: Staff quarter; 17: Meteorological observatory; 18: Resting shed; 19: Sentry-box; 20: Weigh-bridge; 21: Sales counter; 22: Store and canteen; 23: Manure pit; 24: Nursery; 25: Watch-tower; 26: Guide-post; 27: Statue, fountain *etc.*; 28: Flower garden; 29. Cropping plot).

Chapter 12

Layout in Cropping Plots

12.1 Principles

When the cropping plots within the orchard premises have been made ready and the orchardist is ascertained as to the correct planting distance to be maintained in setting up the propagating materials, the next step is actual layout operation to be done in the cropping plots. In the layout operation, the orchardist should however, be guided by certain principles, among which the following deserve particular attention.

1. Selection of the System of Layout

There are a number of systems of layout that are followed in planting fruit plants in the orchards and these have been presented under the next Heading, *i.e.*, 12.2. Each of these systems has definite advantage under particular set of conditions but not in all situations. Thus, which system will be most appropriate for a given plot is a matter that should be decided by the orchardist with great judgement.

In fact, a number of considerations should be kept in view in the choice of a system. Among those, the important ones may be stated as botanical species and horticultural variety of the fruit crop to be grown, propagation method employed, climatic condition of the cropping plot, planting distance to be maintained, cultural practices to be adopted, whether inter-crops, filler trees *etc.* are to be grown and if so, whether these are to be cropped in the early life of the orchard or throughout its running condition, mechanization if any is to be done, availability of skilled labourers and so on.

It should be remembered that unintelligent selection of the layout system may prove to be a serious bottleneck in orcharding with desired success. Also it is not to be forgotten that once a system has been laid out, it cannot be changed to another system in later years when the trees have sufficiently grown. Hence, it should be remembered that "Decision is the horse and action is the cart."

2. Limiting Tree Parts within the Orchard

An important aspect to which the orchardist should not close his eyes in, no part of any tree grown in the orchard such as a branch, an inflorescence, a fruit or even a leaf should cross the boundary of the orchard not only in young stage but also when it has fully grown up. He should note that if any part of any tree growing in the orchard premises spreads out to appear over the property that belongs to other person, it amounts to *intrusion*. Such a protrusion by a tree is a blameworthy act not to the tree but to its owner, besides the fact that this issue may become a source of conflict with the neighbours. The situation could be averted by taking suitable care when planting of the propagating materials is to be done near the boundary of the orchard as has been explained below.

It has been mentioned in the Chapter 14 that the sidewise spread of the crown of a fruit tree assumes more or less a circular appearance when it is fully grown up. The trunk of the tree obviously remains in the centre of the crown, *i.e.*, the canopy. (The same is true with the cluster of stems in case of a shrub and with the pseudostem in the banana herb also. However, for the sake of convenience, the term tree is retained in this text). Since the trunk is in the centre of the crown, the distance from the trunk to the crown periphery, *i.e.*, the radius will be *half the distance of the crown diameter*. For example, if the crown attains a diameter of say, 10 metres in grown up condition, the radial distance from the trunk to the periphery will be 5 metres.

Now, if a distance of less than 5 metres is left in planting from the boundary, in grown up condition of the tree, the trunk will stand less than 5 metres away from it. In this case, the crown which will grow upto 5 metres from the trunk will evidently cross the orchard boundary and thereby, this will to spread over the property that does not belong to the same orchardist. This is shown in Figure 12.1.

On the other hand, if the trunk maintains a clear distance of at least 5 metres from the orchard boundary, no part of the crown will cross the boundary. As a matter of fact, it has been pointed out in the preceding Chapter that the planting distance of a fruit tree is determined in consideration to the diametric spread of its crown. That is, theoretically the distance should be the same as with the spread of the crown but practically it should be slightly more than that. Thus, if the crown diameter spreads upto 10 metres, planting distance should be a little more, such as 10.5 or say, 11 metres. Accordingly, if the propagating materials are planted leaving a distance of half the planting distance from the boundary, the crown will remain completely within the enclosure of the orchard and no part of it will actually be set forth outside the boundary.

Figure 12.1: Branches Crossed the Orchard Boundary as the Distance from the Trunk to the Crown Periphery of the Trees is Lesser than the Distance from the Trunk to the Orchard Boundary.

In consequence to the above, this is an essential step that in the layout operation as shown in Figure 12.2, a distance of *half the planting distance must be left vacant between the boundary and any of the tree which is adjacent to it,* in order that infringement of any part of any tree outside the orchard boundary could be avoided.

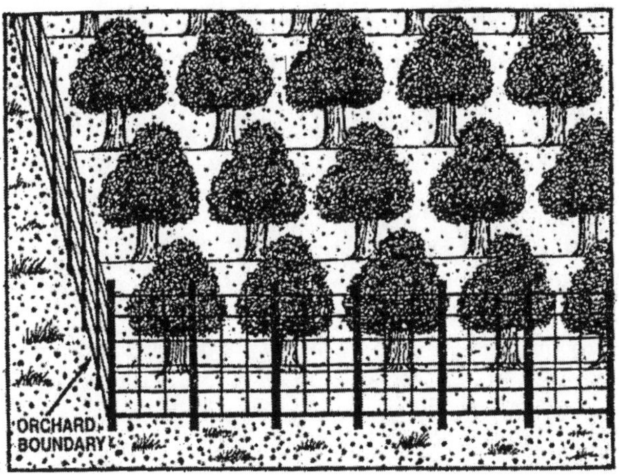

Figure 12.2: Branches Remained Inside the Boundary as at Least Half the Planting Distance has been Left Vacant from the Boundary to each Tree.

It should further be kept in mind that if a space of more than half the planting distance remains vacant from the boundary, then also no planting should be done there. In other words, *at least half the planting distance* should always remain unplanted from any of the boundary line and this is obligatory.

3. Admitting Optimum Sunlight and Air to the Trees

Whichever system of layout that is adopted, the orchardist should be careful in locating the rows of fruit trees in such a manner that they are not much shaded by the neighbouring trees and are not otherwise deprived in receiving optimum amount of sunlight and air. This should not only happen in the early stage of their growth but also when they are in full maturity. Alignment of rows in east-west direction is therefore, a desirable approach to enable the trees to get adequate sunlight in the morning, day time and in the afternoon.

4. Correctness in Layout

A factor of vital importance in the layout operation is maintaining correctness in it. An error committed in the layout operation will lead to planting of propagating materials in improper position. As a result, the layout will not conform to the system and cropping of the fruit trees under the desired system of layout will not be possible. In fact, it is though easy to draw a layout system in black and white, laying out correctly in the actual field becomes difficult unless the persons doing the work are extremely careful at every step.

The materials that are required for the layout operation are pegs, rope, lime dust and mallet. The pegs should be 40-60 cm long, 0.8-2 cm thick, straight and preferably with heads. It is better if such type of iron rods or bars are used for the purpose and in absence, wooden pegs or those made even with bamboo slices are satisfactory materials. Needless to say that these should be strong enough as would not break or bend when they are vertically infixed into the ground by hammering on their heads with the mallet. The wooden pegs or bamboo slices if used should be somoothened by rubbing with sand-papers to avoid injury on hands. The rope may be of cotton or plastic material, highly flexible, about 1 cm thick and sufficiently strong such that it would not tear easily if tightly pulled from both ends.

The *pouring of lime dust* for marking lines on the ground is conventionally done with the help of hands, i.e., palms. However, a better practice may be to apply it through a suitable container as shown in Figure 12.3. To make such a container, a 20-25 cm long and 15-20 cm wide can should be taken, the upper end of which should be open or provided with a lid while the bottom is closed. At the bottom, a suitable hole of 5-8 cm diameter should be made. A lid should be suitably fitted with a hinge at the bottom to open or close the hole. Now, lime dust is poured into the container, closing the hole, *i.e.*, the opening at the bottom (The bottom may be tapered).

To draw a straight line with lime dust on the ground, pegs are tied at both ends of the rope. As shown in Figure 12.4 (a), the rope is then tightly stretched at both ends. While it is so held by two persons on both sides, the pegs are firmly infixed into the ground and hammered on their heads so that they are not loosened. The rope should however, be in contact with the ground in straight line. On doing so, the can containing the lime dust should be brought. The person holding the container should walk slowly along the rope, opening the lower lid of it. By this, the lime dust will be poured over the ground and it should be ensured that it falls over the rope only and not at the side. If he stops walking midway for some reason, he should close the opening by the lid to disallow pouring of lime. In this way, when he reaches at the end, the opening is closed. The pegs are pulled up, the rope is removed and the straight line is clearly drawn by lime marking.

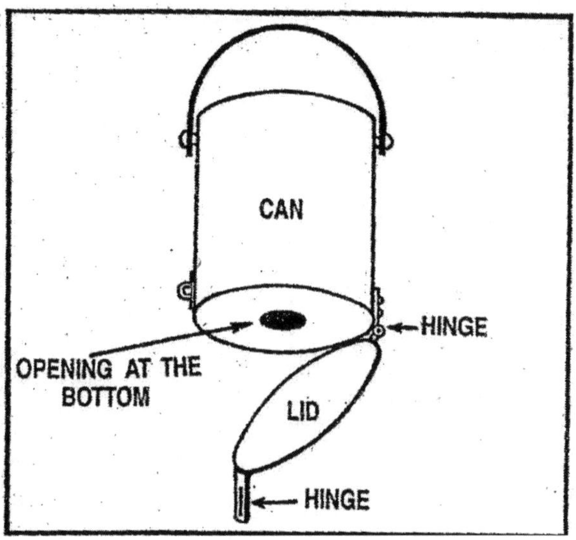

Figure 12.3: A Can with Opening and Lid at the Bottom for taking Lime Dust. (The opening at the bottom may also be made conical).

Now, to draw an arch on the ground, compasses would be required. But so big compasses are not obtainable and hence, a simple type of compasses could be made with rope and sticks. To do so, a stick is tied at one end of the rope while at measured distance at the other end, another stick, preferably an iron rod pointed at one end is tied. The length between these two pegs is therefore, the radius of the arch. As shown in Figure 12.4 (b), to draw the arch with lime dust, the stick tied at one end is firmly fixed at the point which is the centre of the circle, *i.e.*, the arch. The stick or the iron rod tied at the other side with the pointed end downwards is tightly stretched by pulling the rope and at this position, this is slowly rotated making scratch on the ground. While one person rotates it to make the arch by scratching, another person (holding the lime dust container) will walk behind him and pour the lime on the scratch mark through the lower opening of the can. In this way, an arch could be drawn on the ground by lime dust.

It has been mentioned above that to bring about correctness in drawing the layout on the actual plot, utmost care should be taken by the persons carrying out the task. However, incorrectness often appears, such as a line fails to meet at the point where it is desired. In such a case, the operators should patiently trace from the very beginning where the fault has actually occurred. Any attempt to pass through the fault and making manipulation in meeting the line somehow where it does not make so will only invite another fault instead of rectification of what has been committed beforehand. Therefore, tracing from the beginning should not be disregarded to find out the location of the error and getting it corrected. Evidently, to do so, the erroneously drawn lime markings should be thoroughly effaced out at first before the corrected lines are redrawn.

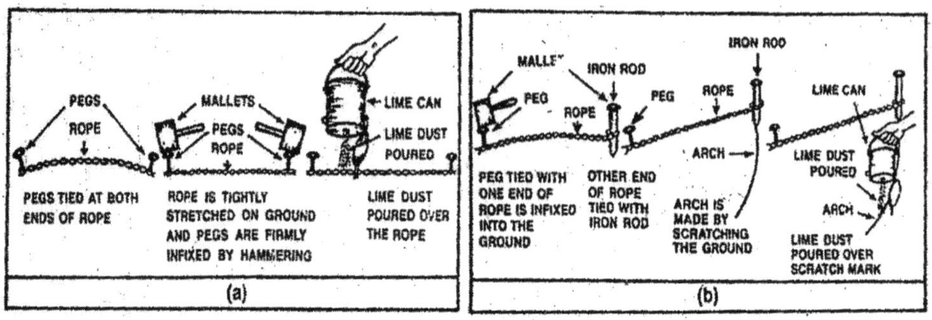

Figure 12.4: Drawing Lines on the Ground by Marking with Lime Dust. (a) Straight line and (b) Arch.

Once the layout has been accurately drawn on the plot by lime marking, pegs should be infixed at the correct positions as early as possible. Delay in this matter should be avoided as the lime markings may become indistinct or even effaced and thus, the actual positions of pegging may not be accurately known. Laying out with lime marking in a cloudy day is hence, not desirable. An important point which should not be ignored is, the pegs should be infixed deep into the ground with firmness by knocking in at their heads with the mallet, in order that they are not loosened apart easily or tilted. For this purpose, embedding pebbles at their bases is desirable.

It should be well-remembered that the orchard may stand in its place for fifty years or more, and hence, hurry-scurry in making the layout and thereby, leaving any error in it will stand in the way in orcharding with success. The entire operation should be carried out with great care and patience.

12.2 Systems of Layout

There are a number of systems of layout of orchard plots that are followed throughout the world. Each of these systems may be considered to be perfect under specific condition only but not in every situation. In the following is described the conventional method of layout by each of these systems, the number of fruit trees that can be accommodated in each such system, if planting distance remains the same and the advantages and disadvantages under each system.

1. Square System

This system is by far, the most commonly followed of all systems. In this system, individual squares are drawn in such a manner that the distance between corners of the squares remains the same as with the planting distance. Planting is done at *each corner of a square* and thus, the distance between rows and between the trees remain the same.

(a) Method of Layout (Square System)

(i) At first a straight baseline AB is drawn. [Figure 12.5(a)] on the cropping plot by marking with lime as has been described above under the Heading 12.1(4). The line should preferably be parallel to a road or border line of an adjacent cropping plot of the same orchard, or the boundary of the orchard premises, if the plot is situated at the edge of it. The point A of this baseline should be near to one end of the plot while the point B should be near to the other end of it.

(ii) A point is selected near the point A on the baseline [12.5(b)]. The point is selected at such a position that if a perpendicular line is drawn over it, a space of at least half the planting distance is maintained at all parts between the perpendicular line to be drawn and the border line of the cropping plot where layout is being done. A peg M is fixed at this point.[3]

(iii) The perpendicular line MD is drawn at the point M where the peg is fixed and over the baseline AB [Figure 12.5(b)]. To draw the perpendicular line, the 3,4,5 unit system may be followed. That is, a triangle is drawn whose base should be 3 parts and altitude 4 parts by length of any unit, or *vice versa*, but the hypotenuse must be 5 parts by length of the same unit in either of the case. Thus, to draw the triangle, a peg Q is fixed on the baseline AB at a distance of 3 metres from the peg M. Taking the peg M as the centre, an arch with a radius of 4 metres is drawn above the line. Similarly, taking the peg Q as the centre, another arch with a radius of 5 metres is drawn in such a way that the latter arch intersects the former.

The arches should be drawn by scratching on the ground with the pointed iron rod or peg tied to the rope and marking with lime as has been described above under the Heading 12.1(4). An important point to be remembered in this connection is, geometric error very likely occurs in drawing the arches under field-scale. The chance of such error is however, less if arches of *relatively long radiuses* are drawn, *i.e.*, by taking greater multiples of the 3,4,5 units, *e.g.*, a base of 6 (= 3 x 2) metres, altitude of 8 (= 4 x 2) meters and hypotenuse of 10(= 5 x 2) metres, or even of still greater lengths.

[3] The purpose of maintaining a clearance of at least half the planting distance from any part of the border lines is to make all parts of the trees confined within the cropping plot. Thereby, any part of any tree is prevented to cross the border lines when it is fully grown up. Evidently, this is of particular significance if the border line of the cropping plot where the layout is being done is the same as the boundary line of the orchard premises. Even if the border line is not the same as the boundary line, it is advisable that no part of any tree should cross the border line of the particular cropping plot. It may be noted that the procedure is to be followed not only in the square system but in all the systems of layout also which have been laid down later.

A peg R is fixed at the point of intersection of the two arches. On doing so, a straight line MD is drawn with rope and lime. In drawing the perpendicular line, utmost precaution sould be taken to ensure that the rope *touches* the peg R but remains freely straight and tight *without forming even a slightest bend* due to any possible writhe caused by the peg P. The point D should be located closed to the other end of the plot and similarly, the perpendicular line DM is projected straight upto the point C, *i.e.*, near the other end of the plot.

(iv) After drawing the perpendicular line, the pegs Q and R which are of no more use should be removed. Then starting from the peg M, planting distances are measured and pegs are fixed on the lines MB and MD [Figure 12.5(c)]. In this way, pegging is continued at measured planting distances on these two lines along their entire length, leaving at least half the planting distance from the border line. The MC part of the perpendicular line should also be pegged at measured planting distances, if the length permits so.

(v) It needs pointing out in this connection that to maintain geometric accuracy and especially when the orchard area is very large, it is advisable that one or more perpendicular lines should be drawn over some points on the lines MB and MD [Figure 12.5(d)] following the same 3,5,5 unit system. The perpendicular lines drawn should be extended near upto the respective border lines. On these perpendicular lines also, pegging should be done at measured planting distances, maintaining a gap of at least half the planting distance from the border lines.

(vi) As shown in Figure 12.5 (e), towards D end on the line, the peg N is selected such that the perpendicular line when drawn over it, a clearance of at least half the planting distance is maintained at all parts between the perpendicular line to be drawn and the opposite border line of the plot. The perpendicular line F is drawn on it, *i.e.*, over the line MN, following the 3,4,5, unit system. The line should be projected backward also upto a point E, which is near the border line of the plot. Starting from the peg N, the pegs are fixed as before on the lines NF and NE at measured planting distances, leaving a gap of at least half the planting distance from the border lines at both sides.

(vii) The point O on the line NF where the peg is fixed and which maintains a clearance of at least half the planting distance from the border line is now joined with the *vis-à-vis* peg P fixed on the line AB [Figure 12.5(f)] at the opposite side with a straight line drawn with rope and lime. The line is extended both sides upto the point G and H, *i.e.*, near upto the respective border lines. Obviously, this line should be perpendicular to the lines EF as well as AB.

If no geometric error has been committed anywhere, the line OP must be exactly perpendicular to the lines EF and AB. But to make sure that no such error has been done in the drawing, it is advisable that a perpendicular line should again be drawn at the point O over the line EF by 3,4,5 unit system [Figure 12.5(f)]. If this drawn perpendicular line meets exactly at the point, *i.e.*,the peg P on the baseline AB, it is a confirmatory proof that no error has taken place anywhere. On the other hand, if the line fails to meet the point P, it points out that error has been committed somewhere.

If any geometric error has at all taken place, tracing should be done from the very beginning to detect and rectify it as stated earlier under the Heading 12.1, *instead of attempting manipulation to meet the line somehow at the proper position which will only make it more erroneous.*

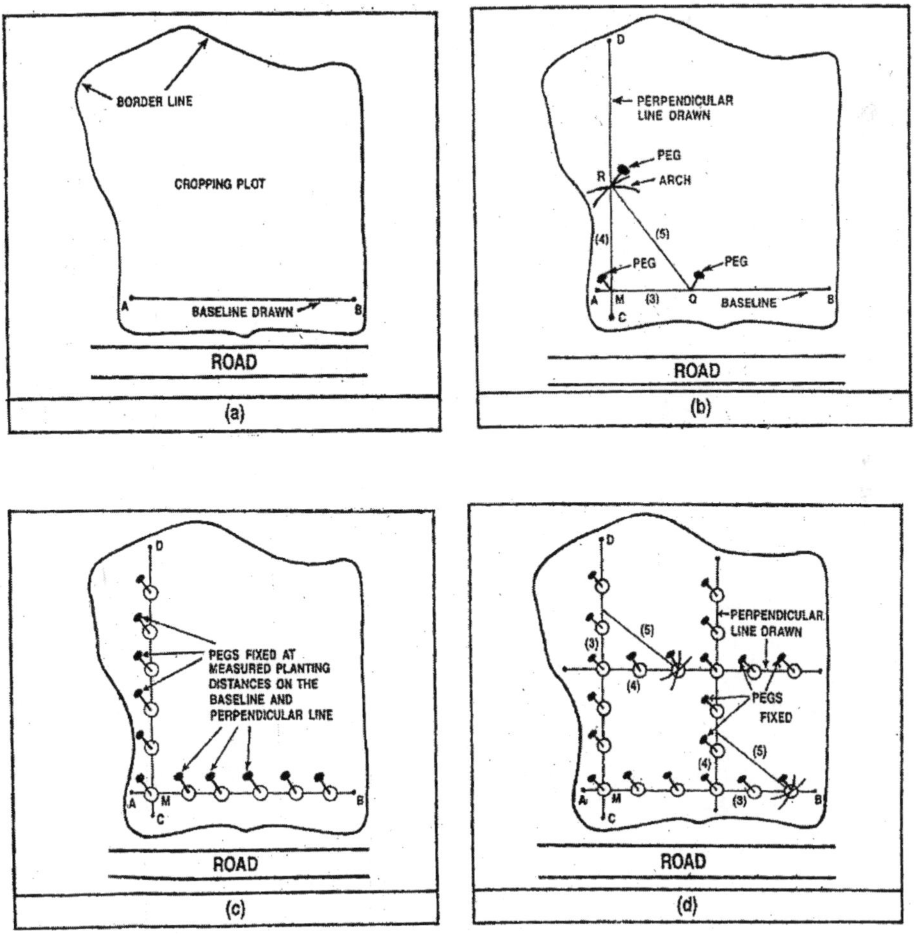

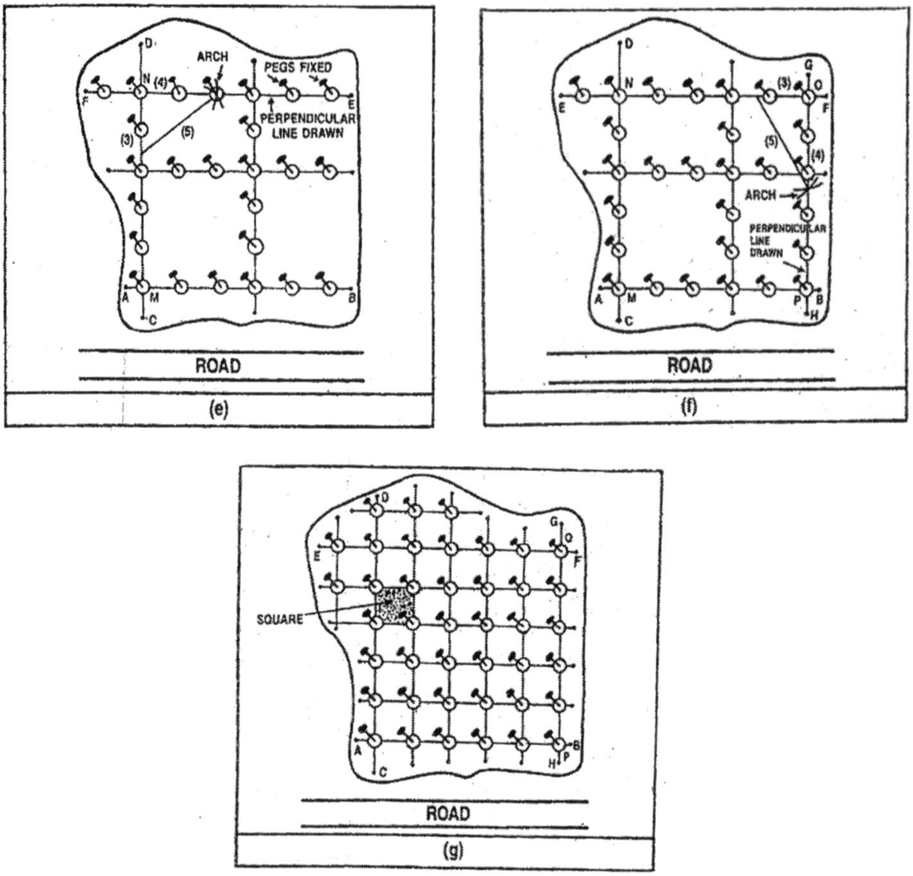

Figure 12.5: Square System of Layout through Steps (a) to (g).

Pegs are fixed at measured planting distances on the line OP. On the extended parts of the line OP also at both sides, *i.e.*, OG and PH, pegging is done leaving at least half the planting distance from the border lines, if the lengths permit so. A large square MNOP is thus, drawn on the cropping plot [Figure 12.5(f)], each line of which has been pegged at equal spaces which are the planting distances. Besides, pegging has also been done on the extended parts of the sides of the MNOP square upto the border lines, where possible.

(viii) Now, as shown in [Figure 12.5(g)] each peg that has been fixed on each side of the large square MNOP is joined straight with its *vis-à-vis* peg, which has been fixed on the opposite side of the square with the help of rope and lime. All these joined lines are also projected further outside the MNOP square to reach near upto the respective border lines of the plot. By drawing these straight lines, many small squares have been formed within the large square, MNOP and may also be outside it depending on the shape of the plot.

Just after making the layout in correct manner, pegs are fixed at all corners of the individual squares within the MNOP square, carefully as described above under the Heading 12.1 (iv). Pegging should also be done at the measured planting distances on the extended parts of the lines outside the MNOP square where it is permissible and leaving a clearance of at least half the planting distance from any of the border line of the plot for reason as stated above.

Lastly, it should not be forgotten that those pegs which were fixed for convenience in drawing and do not indicate the planting position should all be pulled up and removed to avoid any confusion. Thus, only those pegs which are indicators of planting position will remain infixed in the cropping plot.

The pegs remaining infixed should not be disturbed until the propagating materials of fruit trees are planted at those positions. On doing so, they will evidently remain equidistant from one another.

(b) Number of fruit trees to be accommodated

In this system of layout, the row-to-row and tree-to-tree distances remain the same. Accordingly, each tree is provided with an area which is the square of the planting distance, *i.e.*, the multiple of row-to-row and tree-to-tree distances. The number of trees per unit area would thus, be determined as follows:

$$\text{Number of Trees} = \frac{\text{Total Area copped}}{\text{(Row-to-row) x (Plant-to-plant) Distancess}}$$

For example, if the total cropped area is 1 hectare, row and tree distances are respectively 10 meteres and 10 metres, number of trees will be, 1 hectare (=10,000 square metres) ÷ 10 x 10 metres, or, 100.

(c) Advantages and Disadvantages

The following advantages are particularly derived in square system of layout.

(i) Equal space is made available to each fruit tree in the plot.

(ii) Cultural operations and harvesting can be done in both directions.

(iii) Layout by this system can be done without much difficulty.

Among disadvantage, as shown in Figure 12.11, space between the four trees in a square remain unutilized.

2. Diagonal (Syn. Quincunx) System

This system is basically the same as square system, except that an extra fruit plant, which may be a tree, or a shrub, or a herb is planted in the *middlemost position of each individual square* drawn. The fruit plant which in fact, fills the middle part of the square is known as a *filler tree* or a *filter plant* as the case may be, or simply a *filler*. Conventionally, the following conditions are usually taken into consideration in choosing a plant for growing as a filler.

(i) It must be such a tree, or a shrub, or a herb that produces horticulturally recognized fruits (*vide* Chapter 1)

(ii) However, such fruit crops which are of very low height, in other words which do not attain a height of more than 1 or 1.5 metres, *e.g.*, pineapple, strawberry, cape-gooseberry, melons *etc.*, if grown in the inter-spaces of fruit trees are not regarded as fillers. They are in fact, recognized as inter-crops. (For terminology, Heading, 19.1 under the Chapter 19 may be referred to).

(iii) The botanical species to which the filler tree or plant belongs should not be the same as are planted at the corner of the squares.

(iv) Either one or more than one species, or variety of fruit plant may be grown as filler in the same cropping plot.

(v) In grown up condition, it should assume an appearance that should be relatively dwarf and much less spreading than the main trees planted at corners of each square.

(vi) As far as possible, a filler should be *quick-growing* in nature, which means that it should attain maturity to bear economically important fruits within a much short period and usually within 1 or 2 years.

In fact, quite a number of fruit crops may be grown as fillers. Under Indian condition however, the fruit trees or shrubs or herb which are more commonly selected for the purpose are papaya, banana, phalsa *etc.* and those that are less commonly planted include aonla, karonda,ber, custard apple, pomegranate, peach, some temperate berries *etc.* It is pointed out in this connection that in general, the filler trees or plants are grown for temporary period only and hence, these are also referred to as temporary fruit plants while those planted at the corners of the squares are cropped permanently, for which they are termed as permanent fruit trees.

(a) Method of Layout (Diagonal System)

(i) To layout by this system, the square system of layout should at first be drawn to the cropping plot by marking with lime, exactly in the same manner as described above under the Heading, 12.2 (1-a). As length of the side of a square is the planting distance, pegs are fixed at the corners of the square like what has been done in the square system [Figure 12.5 (a-g].

(ii) Thereafter, the middlemost position of each individual square is determined. This can be easily done on finding out the point where two diagonal lines drawn by joining the opposite corners of each square intersect each other [Figure 12.6 (a)].

(iii) The point of intersection of the two diagonal lines respectively within each square which indicates the middlemost position of it is now pegged [Figure 12.6 (b)]. Thus, the four pegs that have been fixed at four corners of each square represent the position where the *main, i.e., permanent* fruit trees are to be planted, while the fifth peg fixed in the midst of the square indicates the planting position of the filler. In order that the *fifth peg* is distinguishable from those that have been fixed at the corners of the square, it is advisable to have some distinguishing mark in the former such as difference in height, colour of peg, shape or colour of its head and the like.

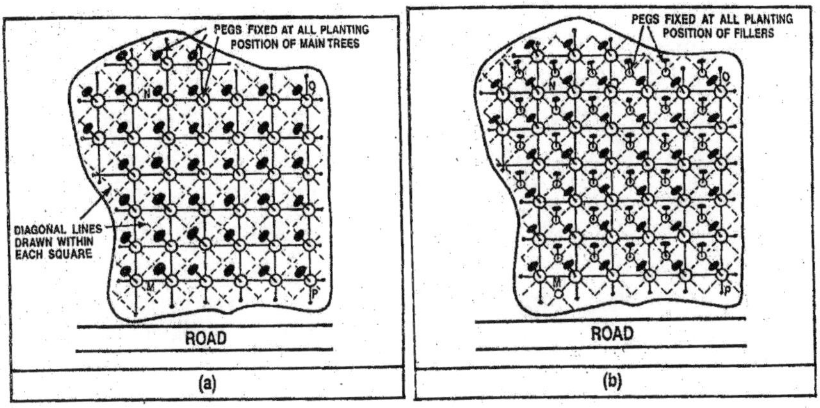

Figure 12.6: Diagonal (Syn. Quincunx) System of Layout through Steps (a) and (b).

(iv) To ensure that all parts of the main trees as well as the fillers remain confined within the cropping plot and they do not send forth any part of them outside the border lines of the cropping plot in their grown up condition, it is necessary that a space of *at least half the planting distance* should be left unpegged from any of the border line. This means that no planting is to be done in that part as referred to under the Headings, 12.1 (2) and 12.2 (1-a-ii).

(b) Number of Fruit Trees and Fillers to be Accommodated

Both for the main trees as well as for the fillers, the distances between rows and between trees or plants in the rows remain same in this system. Thus, each main tree and also each filler is provided with an area which equals to the multiple of row and tree or plant distances. Accordingly, the number of main trees per unit area would be the same as with the number of fillers and these would be determined as follows:

$$\frac{\text{Total Area Cropped}}{\text{Distances Between Rows x Trees (or Plants)}}$$

Although theoritically, the number of main trees and number of fillers remain the same per unit area of land, practically a little difference may be there, if the total number of main trees as well as fillers that have been planted in the entire cropping plot are considered as a whole. The difference is due to the actual number of trees or plants of the cropping plot.

(c) Advantages and Disadvantages

The diagonal system offers the following extra advantages over the square system.

(i) As shown in Figure 12.11, it ensures more utilization of the orchard land and thereby, makes provision for *almost double* the number of fruit trees or plants as compared to the square system.

(ii) More than one species of fruit crop is actually grown in the same cropping plot. Owing to that, return from the orchard becomes higher.

(iii) The cropping of more than one species of fruit crop in the plot gives rise to availability of produce in more than one season. Thereby, any possible loss from one crop that may result in some years may be compensated from the other crop.

(iv) Weed growth become much less as the middle position of each square remains occupied by the filler and not exposed unilike in the square system.

However, the system has disadvantages also at least in respect of the following.

(i) The space between the main trees and the fillers is not adequate in most cases. As a result, overlapping of branches of the neighboring trees may take place, for which performance will be poor for one or both of the fruit crops.

(ii) Due to crowding, acute competition results between the main tree and the fillers in repect of moisture and nutrients taken up from the soil.

Some orchardists are not in favour of this system. They are of the opinion that the filler trees always make competition with the main fruit trees in the orchard. According to them, if filler trees, shrubs or herbs are at all to be grown, these should be uprooted when the main fruit trees would come to bearing condition.

To avoid competition however, somewhat greater planting distance for the main trees is recommended by many. But by such planting, the number of main trees that can be accommodated in the cropping plot become less and thus, the purpose of growing filler trees or plants is not properly served.

As a matter of fact, if cropping of both the main fruit trees and the fillers is if to be accomplished with desired success, suitable arrangements become necessary that the filler trees or plants *do not make much vegorous growth*. Some of the measures to make so have been put down below.

(i) It may be noted that dwarf and less spreading varieties of the species of those fruit crops which are to be grown as fillers are especially suitable than the vigorously growing varieties. Such varieties of some fruit crops have been mentioned in the Chapter 14.

(ii) Proper training operations done from early stage of growth makes the trees dwarf and less vigorous. Pruning should also be done more heavily for this purpose. Especially when phalsa, peach, custard apple, ber, aonla, *etc.* are grown as fillers, drastic pruning should be done to restrict their vegetative growth.

(iii) Regular de-suckering should be a must, if banana is grown as a filler plant. Unwanted suckers should regularly be removed in order that the mother sucker or the main fruit trees are not deprived in getting nutrients and water.

(iv) Drip irrigation could be said to be very ideal if filler trees or plants are cropped. This system checks excessive growth of crop plants and at the same time, enhances productiveness, besides rendering other benefits. Hence, irrigating both the main trees as well as the fillers by dripping could be regarded to be an effective measure for successful culture of all the fruits grown in the cropping plot.

(v) Restricking prolific vegetative growth of the fillers may also be brought about by foliar application of plant growth retardants as stated under the Heading, 14.2 (15) in the Chapter 14. But these growth regulating chemicals should be used with great precaution in respect of concentrations, stage of growth of the plants and their vigour.

It needs pointing out in this connection that dwarfness of the fillers may also act in lowering their productivity performance, either quantitatively or qualitatively or qualitatively or both. To answer this query, it is worthy to add that the main fruit trees that are planted at the corners of the squares are in fact, the *primary fruit crops* while the filler trees or plants that are grown are only the *secondary* or the *supplementary fruit crops*. Hence, the primary fruit crops should never suffer for the sake of the secondary fruit crops, despite the fact that both are grown in the same cropping plot. Primary fruit crops evidently need first priority in every case..

3. Rectangular System

In this system, rectangles are drawn in the cropping plots and fruit trees are planted at *each corner of a rectangle*. That is, the distance between the trees in each row remain the same as with the planting distance but the distance between rows becomes greater than the planting distance. Accordingly, the longer sides of the rectangles form the rows while the shorter sides represent the trees in the rows.

(a) Method of Layout (Rectangular system)

(i) The baseline AB is at first drawn following the same method as in square system. Near the point A, a point is selected on the baseline such that if a perpendicular line is drawn over it, a space of at least half the planting distance is maintained at all parts between the perpendicular line to be drawn and the border line of the plot. A peg M is fixed at this point.

(ii) A perpendicular line MD is drawn on the point M following the same method as stated in the square system. The line is extended upto the point C.

(iii) Then starting from the peg M, planting distances are measured and pegs are fixed either on the baseline or the perpendicular line as desired. That is, if rows are to be aligned along the baseline and parallel to that, pegs are to be fixed at measured planting distances on the perpendicular line as shown in Figure 12.7. On the other hand, if rows are desired on the perpendicular line and parallel to this, pegging at measured planting distance should be done on the baseline.

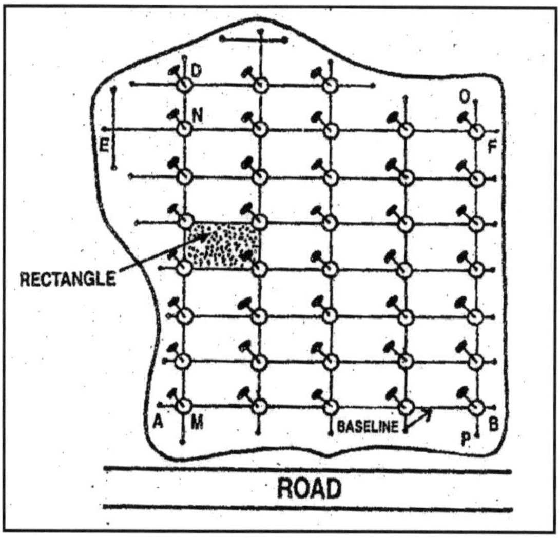

Figure 12.7: Rectangular System of Layout.

(iv) Towards D end of the perpendicular line, a peg N is selected and again a perpendicular line NF is drawn as in square system. The line is extended upto the point E near the border line. Rest of the procedure will be similar to the square system and when the confronting pegs are joined with straight lines, each individual rectangle is drawn in the plot.

As shown in Figure 12.7, trees are planted at *each corner of the rectangle* drawn.

In order that no part of any tree crosses the border of the plot in its mature condition, pegging should not be done within at least half the planting distance from any of the border line as mentioned under the Headings, 12.1(2) and 12.2.

(b) Number of Fruit Trees to be Accommodated

The area provided to each tree is the distance between rows and trees. The number of trees per unit area would therefore, be as follows,

$$\frac{\text{Total Area Cropped}}{\text{Distances Between Rows and Trees}}$$

(c) Advantage and Disadvantage

Wide space left between the rows as is done in this system of layout no doubt, provides opportunity for inter-cropping in the plot in a comprehensive manner. Hence, many fruit growers in different parts of India prefer this system. It is noticeable that this system is usually followed in the cropping of some selected fruit crops, *e.g.*, papaya, banana, aonla, date, phalsa, custard apple, sour lime, lemon *etc.* and not for mango or other fruit trees which are of vigorous growth habit.

If inter-cropping is not done, layout by this system makes a great loss of orchard land as would be evident from the Figure 12.11. However, whether it would be a horticulturally sound practice to cultivate other crops so extensively in between the rows of fruit trees and thereby, lessening their population, or the practice of their cropping in separate blocks unmixed with fruit crops also deserves experimental support under different edapho-climatic situations.

Hedge-row system

This system is basically the same as the rectangular system of layout. The spacing between the fruit trees or shrubs in the rows is usually kept *much lesser* so as to look like a row of compact hedge. It is also the practice in many parts to support the trees or the shrubs in the rows. This is done by fixing bamboo poles at suitable intervals to which are tied the trees or shrubs along with horizontally placed bamboo slices, or with wire-trellises *etc.* Under specific conditions, the system evidently seeks justification. Fruit plants which are dense and of vigorous growth are preferred to make hedge rows to impart compact appearance.

4. Triangular (Syn. Isosceles triangular) System

In this system, fruit trees are planted at the *corners of such isosceles triangles whose base and altitudes are of same length.* As shown in Figure 12.8, when six such triangles laid out in three contiguous rows are viewed together such that one corner of each triangle meets at the same point, a hexagon is formed, but *all the sides and angles of this hexagon are not of same length and degree respectively.* Six trees are thus, planted at the six corners of this hexagon and the seventh tree at the mid-point of this.

(a) Method of Layout (Triangular system)

(i) To start with, exactly the same method of layout as has been described under the Heading. 12.2 (1-a) is drawn. Each individual square is thereby obtained, the length of whose sides represents the planting distance.

(ii) When the individual squares have been drawn by lime marking, at the first row, pegs are fixed at all the cross-points, *i.e.*, at the two corners of each square as shown in Figure 12.8.

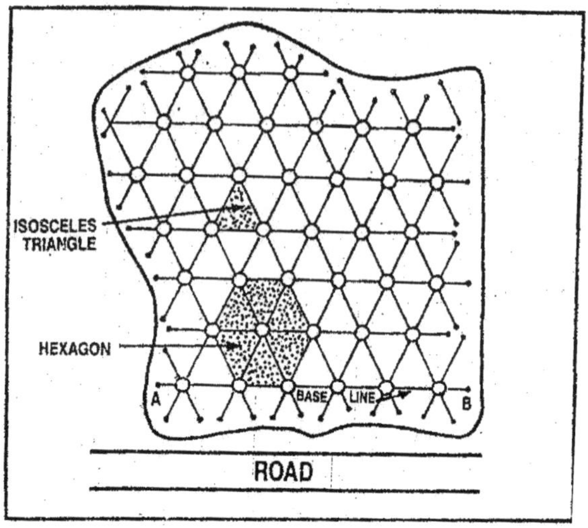

Figure 12.8: Triangular System of Layout.

(iii) At the second row, instead of fixing two pegs at the two corners of each square as in square system, one peg is fixed just at the *middle point of the two corners of each square.*

(iv) Then at the third row, pegging is done in the same manner as has been done in the first row. Similarly, at the fourth row, pegging is done in the same manner as in case of second row. Following this pattern, pegging should be done on all the subsequent rows and continued in the same way upto the last row in the plot. Trees are thus, planted at each corner of isosceles triangles, *i.e.*, whose bases and altitudes are of same length (Figure 12.8).

To discourage any part of any tree to cross the border of the plot in grown up condition, a distance of least half the planting distance should always be left unpegged from any of the border line of the plot. Accordingly, no planting should be done in the unpegged area as has been described earlier under the Headings, 12.1(2) and 12.2 (1-a-ii).

(b) Number of Fruit Trees to be Accommodated

This system is the same as the square system except that in the alternate rows, trees are planted at the *mid-point of the two corners of the squares*, unlike planting done at the corners of the squares as in square system. The number of trees per unit area of land in this system is therefore, almost the same, or slightly lesser than the square system when the distance between rows remain the same in both these systems.

(c) Advantage and Disadvantage

What real advantage has been there in this system of layout is in fact, not very clear. Nevertheless, if the vigour of the trees is very high and the crowns spread over a large area as is noticeable in many trees grown in highly tropical situation, this system may have some benefit. In such situation, overlapping of branches by the neighbouring trees could be avoided even by maintaining somewhat lesser planting distance by adopting this system of layout. The cultural operations, harvesting *etc.* are not performed so easily in this system of layout as compared to the square system. In fact, this system could not be said to be of wide application for the fruit crops grown under Indian condition.

5. Hexagonal (Syn. Equilateral triangular System)

Fruit trees are planted in this system at the *corner of equilateral triangles.* Accordingly, as shown in Figure 12.9 (e), when one corner of six such equilateral triangles laid out in three adjacent rows merge together at one point, they form such a *hexagon whose all six sides are of equal length and thereupon, all the six angles are of same degree.* In this equiangular hexagon, six fruit trees are thus, planted at six corners and the seventh tree in the middlemost position. All the seven trees in the hexagon are also equidistant from one another. The system goes by the term as *hexagonal* system, or the *equilateral triangular* system, to differentiate it from the triangular, *i.e.*, isosceles triangular system as described above.

Triangular System	*Hexagonal System*
Triangles are isosceles.	Triangles are equilateral.
Only the opposite sides and opposite angles in the hexagon are equal.	All the sides and all the angles in the hexagon are of same length and degree respectively
In each triangle, the altitude being of same length with the base, distance between rows and trees are same.	Each triangle being equilateral, the base is of greater length than the altitude. This makes lesser distance between rows than the trees in each row.

(a) Method of Layout (Hexagonal System)

Many orchardists avoid this system of layout for the fear of its difficulty in layout. Although it is true that unlike other systems of layout, this system has some complexty, the author is of the opinion that layout by this system is not difficult to be carried out by them who have patience and take utmost care in geometric drawings and believe that " haste makes waste".

Most important aspect in this system lies in the fact that no perpendicular line is drawn anywhere in this system of layout. Instead, many large *equilateral triangles are necessary to be drawn on the cropping plot* by lime markings. The general procedure to layout by this system is described in the following.

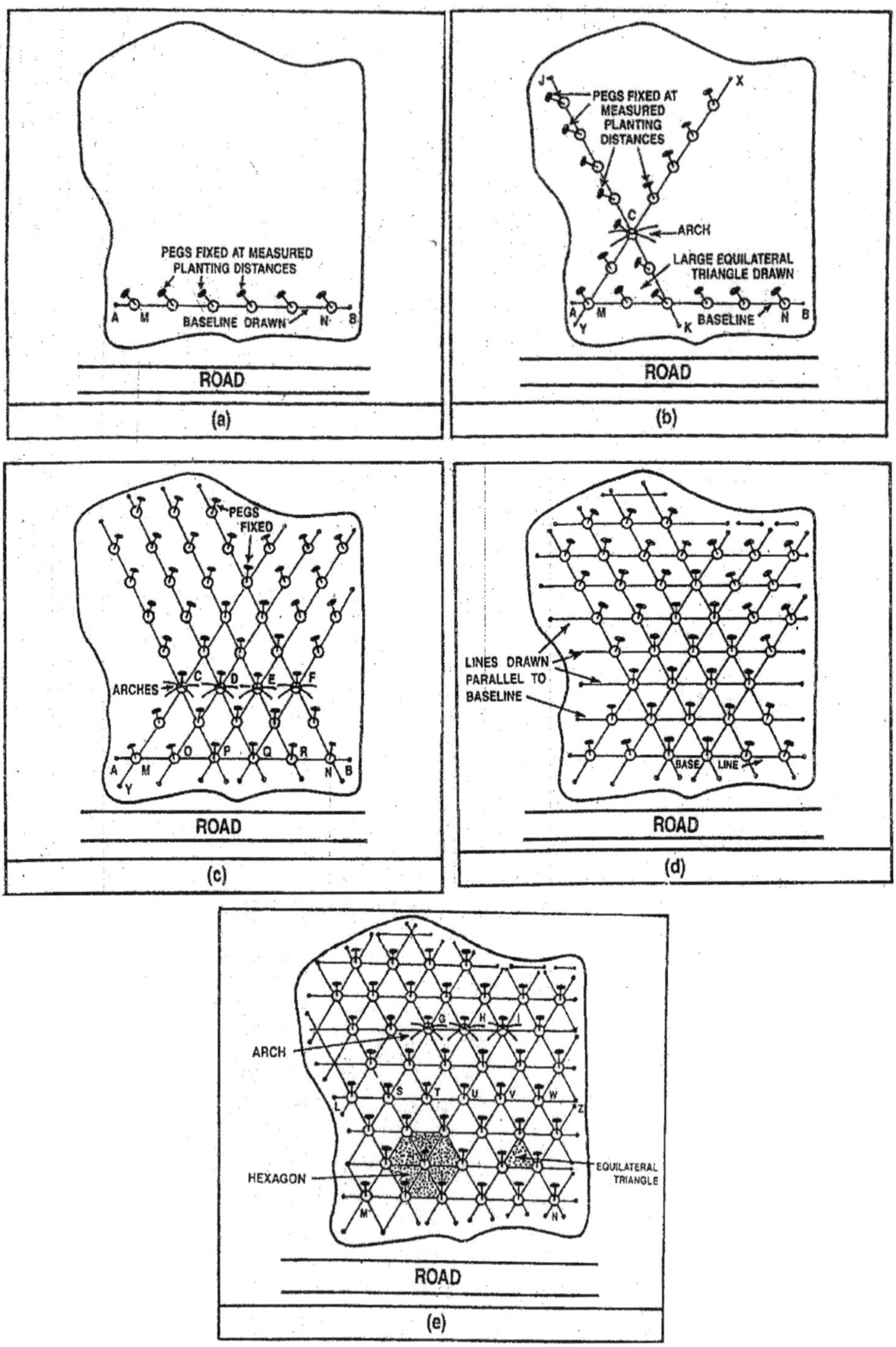

Figure 12.9(a-e): Hexagonal System of Layout through Steps (a) to (e).

(i) As shown in Figure 12.9 (a), at first a baseline AB is drawn on the cropping plot in the same way as in other systems stated earlier.

(ii) A point M is selected near the point A on the baseline leaving at least half the planting distance from the border line of the cropping plot as has been done in the other systems. A peg is fixed at the point M.

(iii) Now, starting from the peg M, pegs are fixed on the baseline at measured planting distance. Pegging is thus, continued near upto the point B such that the last peg N is fixed at least half the planting distance away from the border line as in the other systems.

(iv) After pegging from M to N on the baseline, a peg P is selected on it away from the peg M. As shown in Figure 12.9 (b), taking the length from M to P as the radius, an arch is drawn over the middle part of MP, taking the point M as the centre of the arch, following the method as described above under the Heading, 12.1(4). With the same radius from M to P (= PM), another arch is drawn centering on the peg P such that this arch intersects the former arch. A peg C is fixed at the point of intersection of the two arches.

(v) The points M to C are joined by a straight line. The line is extended both sides upto the points X and Y, *i.e.*, near upto the respective border lines of the plot.

(vi) In the same way as in (v), the points P and C are also joined by a straight line and the line is extended both sides upto the points J and K which are near the border lines respectively.

(vii) Starting from the peg M, pegs are fixed at measured planting distances on the line XY, leaving at least half the planting distance from the border lines at both sides.

(viii) Similarly, pegging is done at measured planting distances on the line JK also, maintaining a space of at least half the planting distance from any of the border line. Thus, a large *equilateral triangle MCP is formed* on the plot, all sides of which remained extended at both sides and that these are pegged at measured planting distances all along their length, leaving clearance of at least half the planting distance from the border lines at both sides.

(ix) Now, as in (iv), other pegs, O, Q and R are selected on the baseline. On doing so, exactly in the same way that have been described above (iv to viii), other large equilateral triangles, which are ODQ, PER and OFN are drawn side by side on the baseline, as shown in Figure 12.9(c). The sides of these triangles are also extended similarly and that these are pegged at measured planting distances, leaving at least half the planting distance unpegged from the respective border lines.

(x) At the next step, as shown in Figure 12.9 (d), straight lines are drawn parallel to the baseline by joining the confronting pegs that have been fixed on the arms of the triangles. The parallel lines are extended both sides near upto the respective border lines.

(xi) Following the same pattern as described so far, another line, say LZ is selected on the plot and on this line also, the points S,T,V and W are selected. As shown in Figure 12.9(e), large equilateral triangles SGU, THV and UIW are again drawn in the same way that have been described from (iv) to (viii). While the arms of the triangles are extended at both ends, the lines are drawn successively parallel to the line CF touching the confroting cross-points. Lastly, pegging is done at all the cross-points, however, leaving a clearance of at least half the planting distance from all the border lines.

As stated earlier under the Headings 12.1(2) and 12.2(1-a-ii), leaving a space of at least half the planting distance from the border lines of the cropping plot, meaning thereby that no planting is to be done there with the idea that no part of any tree crosses the border lines when they are in fully grown up stage.

In the conclusion, it needs pointing out that the method of layout as described above although theoretically appears to be simple and is easy to do so in black and white, the procedure is not so simple practically. This is because, geometric inaccuracy to draw equilateral triangles in field scale with the help of rope, pegs and lime is a very common occurrence. To overcome this error, the following steps should in particular, be taken into consideration.

(i) Equilateral triangles that are to be drawn should be *considerably large* smaller sized triangles usually lead to greater error.

(ii) *More the number of equilateral triangles* that are drawn lesser will be the chance of error.

(iii) Wherever any error is noticeable, the actual point where it has been committed should be detected and there itself is rectified. Any attempt to overlook and somehow manipulating it will only *make the layout more faulty*, without solving the problem.

(iv) It should also be remembered that when any type of drawing is done by a heedful person whether in black and white or in field, not only that he should be extremely attentive in the job, but also he considers it proper to check the correctness at every step in his work before proceeding to the next step.

Use of Set Angles to Draw Equilateral Triangles

The practical difficulty that may arise in drawing equilateral triangles correctly on the plot by rope and lime may be overcome by the use of specially made set angles built with iron rods or bars. The method is described below.

(i) As shown in Figure 12.10 (a), straight iron rods or bars, about 1.5 to 2 metres long and 1 to 1.5 cm thick should be taken.

In each such rod or bar, both ends should be curved so as to make U-shaped hooks. The hooks should be so made that they lie on the same plane, besides that the ends of the hooks at the opposite ends be directed face to face. [Figure 12.10(a)].

The rod or the bar as the case may be is then gradually and carefully bent at the centre to such an extent that even though it curves to become rounded instead of forming any angle, if two lines are drawn along the two arms of the set towards its narrower end, they would meet together forming an exactly 60° angle [Figure 12.10 (a). (N.B. : instead of using iron rods or bars, the set angles may be prepared with wooden bars or bamboo slices also by fitting suitably with nails and rope, but the material should be tough as would not break easily).

The two arms of the set are then conjoined with one or two short pieces of iron rods or bars as the case may be by welding [Figure 12.10 (a)]. This is with the idea that the 60° angle of the set is not disturbed due to rough handling in the field.

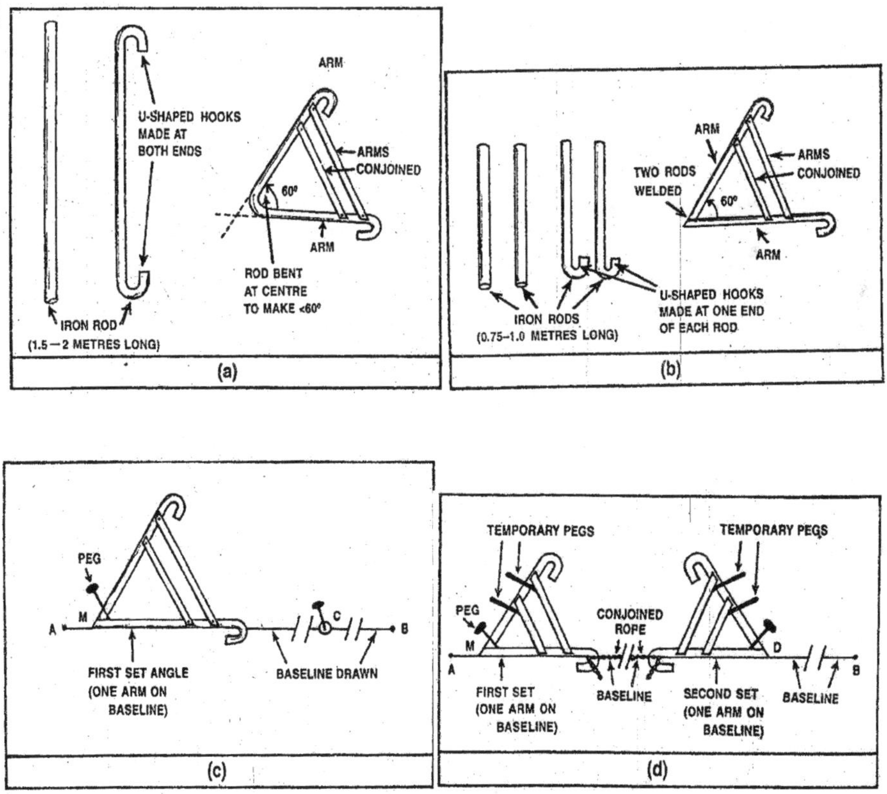

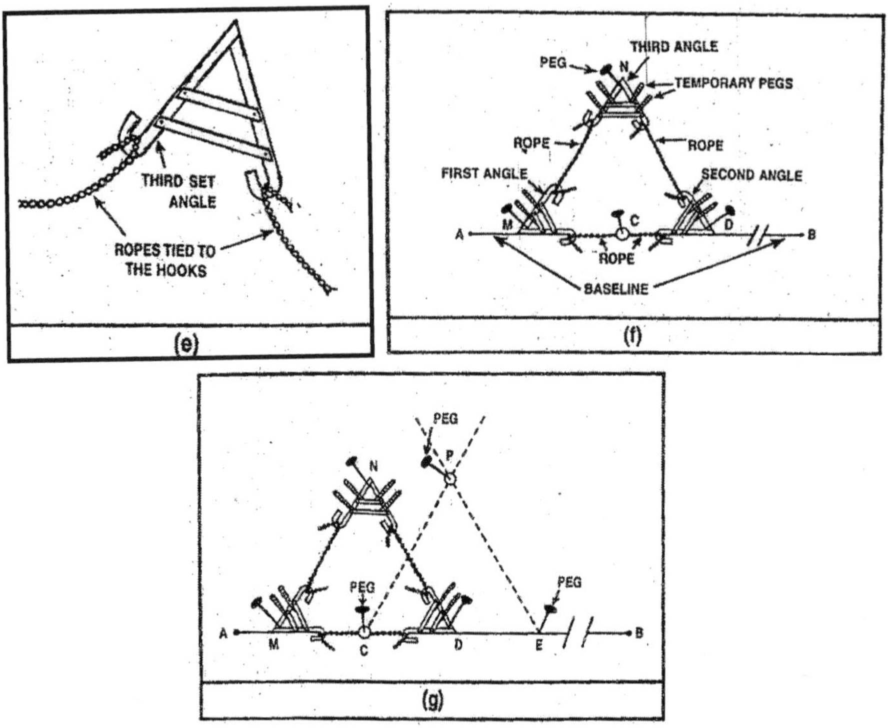

Figure 12.10 (a-g): Equilateral Triangles Drawn with Set Angles, through Steps (a) to (g).

A better method may however, be as shown in Figure 12.10(b). In this method, two separate iron rods or bars, each of about 75 cm long should be taken. A hook should be made at each end of it. The other end of the two should be welded together such that an exactly 60° angle is formed by the two pieces. The two arms of the triangle should then be conjoined by short pieces as above.

It should be noted that to do the layout, three such set angles are needed to be used simultaneously.

(ii) To start with the layout, a baseline AB is at first drawn on the cropping plot [Figure 10.10(c)] following the same method as stated in other systems.

(iii) Near the A end of the baseline AB, a peg M is fixed [Figure 10.10(c)], leaving at least half the planting distance from the border line.

(iv) Now, the first set angle is brought to use. As shown in Figure 10.10(c), it is held in such a position that the peg M fits into the 60° angle of it and the basal arm of it is carefully laid just over the baseline AB.

(v) On doing so, the second set angle is brought to use. As shown in Figure 10.10 (d), it should be placed fitting its notch, *i.e.*, the 60° angle into another peg that has already been fixed on the baseline, say the peg D.

This set should also be placed with care that its basal arm is laid just over the baseline and thus, the 60° angle is faced opposite to the angle set at the peg M.

(vi) The two set angles so placed on the baseline should not be disturbed even slightly from their positions. To ensure that they are not shifted from their set positions, temporary pegs need to be fixed at the notches of the arms which connect the two arms of the angles [Figure 10.10 (d)].

(vii) The basal arms of the two set angles are then made to conjoin together with a strong rope by fastening very tightly at their hooks that tie on the baseline [Figure 12.10 (d)]. By doing so, the basal arms of the angles to which the connecting rope is tied will obviously be stretched over the lime-marked baseline.

(viii) The third angle is now brought to use. To the two arms, *i.e.*, the two hooks of it, two separate ropes are fastened [Figure 12.10 (e)].

(ix) Thereafter, as shown in Figure 10.10 (f), the free ends of these two ropes are fastened respectively to the two hooks at the upper ends of the first and the second angles, taking utmost care in adjusting the length of the two ropes that the total length running from the corner point of the third angle to the peg M and similarly, from the same point to the peg D are exactly the same as with the length from the pegs M to D, when the ropes are tightly stretched on the ground. This may be done by carefully stretching the ropes and the set angle in a tight manner. A peg N is fixed at the corner of the third angle, *i.e.*, at the 60° angle MND. In this angle also temporary pegs are to be fixed at the notches of the connecting rods to avoid its positional deviation.

(x) Thus, as shown in Figure 10.10 (f), a large triangle, MND, which is made up of three set angles tied with rope would be seen falling on the ground whose base is stretched over the lime-marked baseline AD. As the three sides of this triangle, *i.e.*, MN, DN and MD as well as three inside angles,i.e., MND, NDM and DMN are respectively of same length and degree, the triangle is evidently an equilateral one.

Lime dust is now poured over the ropes including the arms of the three triangles, so that the large equilateral triangle is marked on the ground. The lines along arms of the triangles are extended straight near upto the respective border line with rope, peg and lime. Pegs are then fixed on the lime marked lines at measured planting distance, starting from M to B and M to N and the extended part, however, leaving at least half the plating distance from any of the border line. The pegs indicate the position where planting is to be done.

(xi) The entire set, *i.e.*, the flexible equilateral triangles joined by ropes is then taken out from its position. Following that, the notches of the first and the second sets are fitted with other pegs, say C and E on the baseline AB and the third angle is set keeping the ropes straight. A peg P is fixed at the notch of the third angle and similar lime markings and peggings are done to draw another equilateral triangle as shown in Figure 12.10 (g).

(xii) In this way, as many equilateral triangles as possible should be drawn and the lines are to be extended near upto the border line. Pegging should be done at measured planting distances, leaving a clearance of at least half the planting distance from the border lines.

(b) Number of Fruit Trees to be Accommodated

Lesser wastage of space between the trees and shorter the distance between rows make it possible to accommodate greater number of trees in hexagonal system of layout. Accordingly, when the planting distance between rows and trees remains the same, 15 per cent *more trees could be accommodated in the hexagonal system of layout than the square system.*

(c) Advantage and Disadvantage

Hexagonal system of layout may be regarded to be the *best of all the prevalent systems* for the following reasons.

(i) Wastage of land is least in this system, meaning thereby that the system makes *greater utilization of land than the other systems* of layout. This is also the fact in the diagonal system but in that system, the filler trees or plants that are planted to fill the inter-spaces do not belong to the same species as with the main fruit trees.

Less wastage of space between the fruit trees in the hexagonal system is however, a consequence of the fact that when a tree has grown up, it *assumes a more or less circular appearance of its spread.* A hexagon also conforms to a circle more equiably when it is equiangular and thereby, the circular spread of a tree covers the hexagon more neatly than any other system. This would be evident from the Figure 12.11.

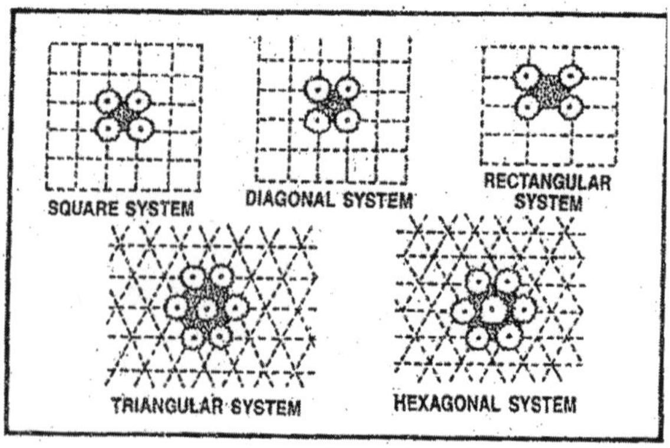

Figure 12.11: Space Left between Fruit Trees in their Grown Up Condition in different Systems of Layout.

(ii) Each tree is provided equal space yet the branches of one tree do not overlap those of the neighbouring ones.

(iii) The cultural practices and harvesting may be done in three directions.

(iv) Weed growth is also less in this system as exposed land between the fruit trees is much less as in diagonal system.

Despite all advantages, the system poses to have the drawback in that, the layout is the field by this system is not very simple and requires *geometric accuracy* at every step.

Unfortunately, the system has not received emphasis by the orchrdists of India and this may be ascribed as due to three reasons. These are firstly, the square and partly, the rectangular systems of layout are followed in this country since long past and the new orchardists are in fact, reluctant to change this age-old practice. Secondly, a large section of the orchardists are in fact, ignorant of this system and thirdly, laying out correctly by this system is difficult to be done by the ordinary orchardists themselves without the help of technical personnels. Technical persons are also not available easily.

6. Free-style System

In this system, planting of fruit trees is done not in any regular geometrically laid out system, which means that the orchardist may plant in *any fashion he likes.*

The system is usually not adopted where commercial orcharding is desired. In other words, this is followed to meet some specific requirements only and the layout is shaped in accordance with that. Some of the situations where this type of planting is done and the method involved are stated below.

(a) Homestead Land

Planting fruit trees in homestead land is most commonly done in the rural than in the urban areas in India. The plantations are laid out keeping in view a number of factors such as requirement of fruits and fuelwood for the family, situation of the land, security, perishability of the fruits and so on. The common practice is to plant more than one species of fruit trees and the selection of the species is done very judiciously.

Such trees whose fruits are regularly required in the family, *e.g.*, limes, lemons, papaya etc. for cooking purpose are planted near the dwelling house or even within the compound and these are usually inter-planted, maintaining close spacings. On the other hand, large sized and vigorously growing trees like mango, jackfruit, jamun *etc.* are planted away from the building to avoid damage of the building by the thick roots of those trees, damage due to breakage of branches in heavy storm and perhaps to avoid, the crows which make nests in the trees and leave excreta.

Where cattle are reared in the homestead land, there is a tendency to plant vigorously growing fruit trees and especially mango, haphazardly around the cattle-shed. The purpose is, to provide shade to the animals and also to stack the hay on the main branches and their angles, besides securing fruits. Where sheep and goats are reared, close planting of jackfruit in somewhat rectangular system of layout is also met with in many parts, with the purpose to feed the animals with leaves of these trees.

Hedge-row of planting with mulberry, phalsa, dwarf and wild ber, karonda, some minor fruit plant species, some berries of temperate regions *etc.* in order to erect brambles is a very common sight in various parts of India for the purpose of security, especially around vegetable gardens. These rows of brambles are also utilized for propping of grapevines where the crop is suitable to be grown.

Close planting in blocks without any regular system is a practice done by many growers for some fruit crops and especially, custard apple to improve pollination. In apple and some other temperate fruits also, trees of the pollenizer varieties are closely interplanted in rows to ensure fruitfulness.

On both sides for the paths or alleys, low-heighted fruit trees or shrubs, notably guava, papaya, pomegranate *etc.* are set in one or two rows maintaining close spacing. Banana suckers are planted in rows where watering could be done easily and is usually at a distance from the dwelling place (which is perhaps because of superstition, prevailing with some people). Tall trees are not planted by many people close to the building or the hutment to prevent lightning.

In addition to the above, there are many other considerations which are kept in view by the rural people of India in planting fruit trees in the homestead land in regard to selection of the fruit trees, their location and the manner of their setting or the layout.

(b) Public Compounds

Planting fruit trees without any orderly system of layout is also done in large compounds around offices, inspection bungalows, circuit houses, police lines, town-halls, courts, schools, colleges *etc.* The common practice is to plant tall trees away from the buildings and the dwarf ones near to those.

(c) Afforestation

Some fruit trees and especially those having timber value such as jackfruit, *Diospyros* species, walnut, plums, filberts, seedling mango, madhuca, plamyra palm *etc.* are sometimes preferred for planting in forests. Formerly, the system was to plant them here and there in a haphazard manner but nowadays, these are usually planted in blocks. Although geometrically laid out systems are not followed due to practical difficulty, as far as possible, planting is done in roughly made square and rectangular systems.

(d) Social and Agroforesty

Planting some fruit trees in these groves is almost a must. The general tendency is to inter-plant different species of fruit trees in rows and around the boundaries.

(e) Mass Planting

Free-style system of layout is followed for planting of fruit trees in large blocks of land in the rural or the urban areas. Fruit trees and other trees are usually inter-planted in these plantations and are set in such a way that the taller ones are located in the centre while the dwarf ones gradually away from it towards the periphery.

(f) Utilization of Wasteland

For productive utilization of inferior land, some fruit trees or shrubs that are hardy, *e.g.*, custard apple, ber, phalsa, guava, jamun, palmyra palm, wild date palm, cashew, elephant apple, *Diospyros* species, pomegranate *etc.* are sometimes preferred. To make the plantation, a number of considerations are however, necessary to be made, such as evenness of the land, its texture, depressed pocket, swampy part, erosion *etc.* and the trees or shrubs are intelligently set in accordance with that. Somewhere it may become necessary to maintain close planting and somewhere at wide distances. Again, roughly laid out hedge-row or triangular systems are followed in some parts to check erosional loss of soil.

(g) Along Avenures, Railways Tracks, Rivers, Streams, Tanks *etc.*

Some fruit trees are sometimes planted at these places. Planting is done in one, two or more rows in alternate manner, similar to triangular system.

(h) Coastal Areas

Large-scale planting of trees is done in the coastal areas among which some fruit trees also find place. Geometrically laid out system is not followed for planting in these areas and free-style system is adopted for practical purpose.

7. Contour System

This is a specialized system of planting fruit trees and is *exclusively adopted* in hilly terrains and high slopes having a *rolling topography* and thereby, having the danger of soil erosion. The objective of this system is to retard soil erosion as much as possible, in order that the slopy land is rendered suitable for growing fruit and also other crops.

Where slopes are steep, necessary ground works like terracing, bunding *etc.* may have to be done at first before planting. The layout for planting should be done judiciously, in order that the roots retard erosion and are not exposed over the ground. For this purpose, rows of the fruit trees or shrubs are laid out as far as possible at right angles to the slopes and that they are planted in an alternate manner between the rows (Figure 12.12). Selection of the fruit trees or shrubs that would be planted in this system is no doubt, a very important consideration. Such type of trees or shrubs which produce dense and highly ramified root system as may bind the soil like claws sould only be chosen.

The contour system of planting has long ago been elaborated by Dutta (Ref. Dutta, S. 1966. *Horticulture in the Eastern Region of India.* Ministry of Food and

Agriculture, Govt. of India, New Delhi) under the eastern hilly area of India where the system has large application. It has been suggested that the contour line should be designed and graded in such away that water flows through the irrigation furrows at a slower rate. Thereby, it is able to penetrate the soil but does not cause erosion.

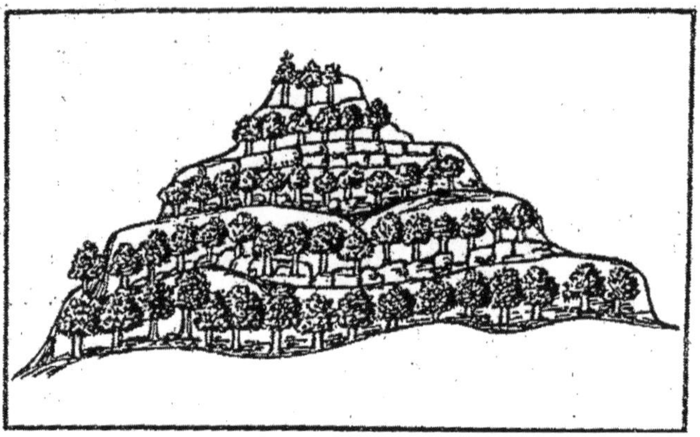

Figure 12.12: Contour System of Layout.

The nature of the slope actually determines the width of the contour terrace. A terrace is regarded to be 'narrow-based' when its width is 12-18 metres and 'broad-based' when more than 18 metres. The chance of erosional loss is lesser with greater the width. The planting distance of the trees or shrubs is usually not uniformly maintained when the slope is greater.

Three types of terraces have been suggested by Dutta that may be adopted in contour planting. These are (i) bench terrace,(ii) graded terrace and (iii) level terrace which are described below.

(i) Bench terrace is designed to control erosion and stabilize the slope so as to make it suitable for cultivation. A step-like structure if formed with ridges on the outer side, along the irrigation grade line. These ridges are planted with fruit trees or shrubs. However, these are also protected by growing some cover crops such as lemon grass or perennial legumes which is of much help in binding the soil and prevent its erosion.

(ii) Graded terrace is designed to intercept and divert the rain water so that it slows down the flow of water. Wide ridges are made in this system and the spacing between the channels is maintained in accordance with intensity of rain, slope of the land and other characters of the soil.

(iii) Level terrace is done usually with provision of embankments and is practised in areas of low rainfall, with the idea to collect the rain water and make it available for the plant that are grown.

Among the three terraces, the *bench terrace is considered to be easiest* to adopt and has greater applicability in hilly terrains.

Chapter 13

Preparation of Cropping Plots

13.1 Necessary Measures

After the cropping plots are laid out, their soil needs meticulous preparation. Preparation of the soil should be done keeping in view the requirement of the crop(s) that is(are) to be grown there and the condition of the soil.

On superficial inspection of a plot, it may seem to the orchardist that it is perfectly suitable to the crop that would be raised there. But many a time, disappointment may be a consequence when it is observed that the crop has either failed there or its performance has not been upto the mark. Thus, instead of forming any theoretical opinion just by visual observation of the land and guessing, the scientific approach is, to have a perusal of the land with the help of standard methods and only on doing so, suitable corrections should be done as are felt necessary. Nevertheless, it needs mentioning that visual observation is also greatly valued when done by a highly experienced person. (Interestingly, many fruit growers and cultivators in different parts of India are met with who are very expert in this work. They tell almost correct result at least for some soil characteristics but usually cannot express clearly how they have detected so).

The steps that need particular attention regarding preparation of that land in the orchard which is meant for growing fruit crop or inter-crop have been brought forward in the following.

1. Making Free from Unwanted Objects

As it has been done in primary operations of the land which has been referred to under the Chapter 7, in preparing the cropping plots also, unwanted materials such as trees, shrubs, tall grasses, broken hutments, buildings *etc.*, if noticed should be removed out. Evidently, this should be done *more scrupulously* before preparation of the cropping plots.

2. Levelling

Similarly, as stated in Chapter 7, the declivity, undulations, earth mounds *etc.*, if found to exist in the plot should be transformed to render it to a flat land by suitable soil shifting process. It should be remembered that although in other parts of the orchard, it is not so essential but for a cropping plot, a flat type of land is absolutely necessary.

3. Soil Testing

Whether the physical characteristics of the soil are favourable to the crops that are to be grown and if not, what measures are to be taken is an important aspect and for this, knowledge of the physical characteristics of it is necessary. In the same way, whether the fertility status of the soil is all right for the crop(s) to be grown and if not, what fertilizers and manures are essential to be known. Accordingly, to have an understanding of the physical and chemical characteristics of the soil of the cropping plot, it is an inevitable task that the soil should be analyzed in a laboratory without which, status picture of the soil is not obtained. Evidently, such analyses are to be done before any manure, fertilizer, pesticide, amendment or other agro-chemical is applied to the soil.

The analyses of the soil should be done, taking *samples* from the cropping plot. It should be remembered in this context that the samples of soil are to be taken from the cropping plot in such a way that it *represents the soil of the entire cropping plot.* In fact, this is an operation that should be done very carefully. For correctness, the samples should be obtained from several parts of the plot and not from one or two sites only. For taking soil samples, a soil-auger, or a pick-axe or a spade may be used.

Another important point to be kept in mind is, the samples should be taken from different depths and not from the surface only. As the roots of fruit trees go to much greater depth than other crops, for establishment of an orchard, it is necessary that the samples are to be obtained from greater depths also, i.e., upto 45 or 50 cm depth.

The samples taken from each depth at different locations should be blended separately and may be dried if these are wet. After this, the representative sample should be prepared following standard procedure, by making several lots and sampling from each such lot. Each of the sample meant for analysis should be taken separately in a plastic or cloth bag and labelled, furnishing adequate informations regarding the sample. It is brought to notice in this connection that in many a case, error is committed in sampling of the soil and its labelling. If it so happens, the analytical report for such a sample has no value.

As many physical and chemical components that are possible to be assayed should be done, however, adopting standard analytical procedure in every case. At least, the reaction of the soil, total soluble solids by electrical conductivity, organic carbon, available amount of phosphorous and potash are the essential components to be considered for analysis.

Soil Amelioration

The reaction of the soil solution, *i.e.*, its *p*H value is beyond doubt, a very important edaphic factor that affects growth and productiveness of a crop and same is true for a fruit crop or an inter-crop, that is grown in the orchard. It is generally observed that majority of the fruit crops do well in neutral, near neutral or in slightly acidic soils. Nevertheless, as mentioned under the Heading, 6 in the Chapter 4, some fruit crops are there which have a preference for such soils that are more acidic in nature and some are found to grow satisfactorily even in somewhat alkaline soils. Such fruit crops are however, not large in number.

Anyhow, it is a fact that the performance of a fruit crop may not be to that extent, if grown in a soil *p*H that is not agreeable to it than when in a soil, the *p*H of which is favourable. Therefore, this is suggestive of the fact that the *p*H of the soil solution should first be known and then *reclaim it, which means to convert it* to the desired *p*H range. This conversion of the soil *p*H is done by application of some substances, known as *soil amendments.*

It is observed that especially in high rainfall area, the soil reaction is acidic due to leaching down of the soluble *bases* in a continued manner. However, to raise the *p*H of acid soil, so as to make it neutral, or even alkaline, lime is the amendment that is universally used. But it is pointed out that the *p*H value of the soil solution indicates the *active acidity* only and not the total acidity, while the bulk of the total acidity is constituted by the other form, *i.e.,* the *potential acidity*, and the amount of lime that is to be applied depends in fact, on the total acidity of the soil solution. The requirement of the amount of lime, necessary to enhance the *p*H upto a specific limit may be determined in considering the relationship between the soil *p*H and the percentage base saturation of it. In general, to raise the soil *p*H from 5.5 which is regarded as strongly acidic to the *p*H of 7.0, *i.e.*, neutral, about 9 and 8 tonnes of limestone may be necessary to be applied per hectare area of land in case of clayee loam soil, respectively in cool, temperate hilly region and warm humid plain lands. This amount of limestone would be lesser by degrees in loams and silty loams, sandy loam and loamy sand respectively.

While acidic soil is usually a feature in high rainfall area as stated above, alkaline soil is a characteristic feature in most of the arid and semi-arid areas as the minerals are not fully leached there. However, other reasons for alkalinity may also be there. The alkalinity occurs when there is a high degree of base saturation. In alkaline soil, that is specifically recognized as *sodic* or *non-saline alkali soil,* the exchangeable sodium percentage may be more than 15 and the *p*H goes more than 8.5. High alkalinity in such soil which is due to sodium carbonate makes discolouration of this soil to look black and hence, the term, *black soil* is also used. The alkaline soil may remain singly or in association with salinity of different degree and thus, two other types or related soils are recognized, which are saline soil and the saline-alkali soil. In case of saline soil, the exchangeable sodium percentage is lower than 15, the *p*H is also below 8.5 and the soluble salts are chlorides and sulphates of sodium, calcium and magnesium. When the soil is both saline and alkaline, it is regarded as *saline-alkali* soil, the exchangeable sodium percentage of which is more than 15 and the *p*H usually is less than 8.5.

To reclaim the alkali soil, *i.e.*, to bring down its *p*H value, soluble calcium salts are usually applied and the gypsum is the most commonly used amendment. These convert the carbonates of bases into sulphates. The gypsum reacts with the sodium carbonate or adsorbed sodium and converts it to sodium sulphate which is leached down. Apart from gypsum, sulphur dust is also very much helpful in soil where sodium carbonate is usually high. When applied, this is oxidized to form suphuric acid and removes the alkalinity like the gypsum, *i.e.*, produces sodium sulphate as the end-product which is leachable. Between gypsum and sulphur, the latter is generally considered more effective in many soils for in the case of gypsum, the carbonate remains present as salts of calcium but when sulphur is applied, the carbonate is totally removed by the formation of carbon dioxide. (Drenching the soil with commercial grade sulphuric acid in highly diluted form is sometimes advocated, instead of sulphur dust. Though costlier, amount of it to be required is much lesser and has an advantage in that, weeds, insect larvae *etc.* are instantaneously killed on application, which however, is to be done with great care). Although alkalinity may be transformed, salinity of the soil appears to be a very tough situation to reclaim. Theoretically, the salts are needed to be leached down in such soil, but practically it is difficult to be done by easy means. Improvement of drainage in the sub-surface, if possible to be done may cause leaching to some extent. Soil with high degree of salinity is therefore, not considered suitable for fruit growing.

To get the desired result, application of an amendment on the soil surface of the cropping plot should be done with great judgement, considering a number of factors and notably the following:

(i) Amount that is to be applied should be adjudged very carefully, considering the *p*H of the soil solution, rainfall, physiography, soil texture, previous crops that were grown, the crops to be raised on the plot of land and other factors. It should be remembered that while under-dose application will not produce desired result, over-dose application will be greatly harmful.

(ii) Application of the amendment should be used in as finely form of particles as possible.

(iii) Instead of bulking at one place, uniform spreading of the amendment should be ensured. For this purpose, it is advisable to work with a spreader which is used for spreading fertilizers.

(iv) Incorporating the amendment material with soil needs to be done very thoroughly. To do so, repeated ploughings with one or two deep ploughing or spading become necessary in most cases. Soil turning ploughs should be preferred to do the operation.

(v) After application, *heavy irrigation should be a must*. This should also be followed by few irrigations more which may not be heavy.

(vi) After irrigation, a definite lag period is no doubt, essential to elapse before planting is done, failing which the planted materials may be damaged or killed.

Fruit Crops Adaptable to different Soil Reaction

Although some fruit crops are able to grow and produce in soils of various reactions, some are very fastidious and they grow satisfactorily with higher productiveness both in terms of yield and quality of fruits when they grow in some soil reaction which is most favaourable to them. In the following has been given the names of some fruit crops which favour acidic and alkaline soils.

Thus, for fruit crops, *viz.*, citruses, apple,grapes, plum, watermelon, blue berries *etc.*, somewhat acidic soil is found to be suitable. Alkaline soil of moderate degree is found to be suitable for fruit crops *viz.*, coconut, banana, guava, date palm, pomegranate, grapes *etc.* Same fruit crops are found to grow in moderately saline soil also.

4. Manuring and Ploughing

It has been stated in the Chapter 8 that where the orchard is large in size, basal application of manure in the orchard land is not necessary because manure in such large orchards are to be applied in the cropping plots only and when the eropping plots would be laid out. On the other hand, when the orchard is not large, the entire area is cropped and roads or other accessories are not provided. Accordingly, in such type or orchards, basal application of manure becomes necessary after primary operation of the land.

Whether basal application of manure is done or omitted, it is advisable to apply at least some amount of *organic manure* in the cropping plots after these have been laid out. This mannurial dose should be in addition to that, which should be applied to the fruit crop as per recommended schedule. (The fruit scientist of the yesteryears, Dr. Adolf Mckenzie remarked, "Fruit trees love organic manure for their anchorage roots and for feeder roots. Apply more and more organic manure for they have no aversion to that. Soil condition and productive capacity of the fruit trees will be markedly increased by high application of such manure."). If manuring is done in the cropping plots, it should be thoroughly mixed with the soil by ploughing and preferably with a soil turning, *i.e.*, mould-board plough. Ploughing should be followed by levelling of the land which may be done by *bullock drawn henga, pata, chowki, moi etc.* or with concrete roller.

13.2 Assignment of Cropping Plots

After the cropping plots have been laid out, it is a necessary part to *assign those with titles* for their easy recognition. Attribution of titles could be done in various ways. Ascription of numbers is an easy system, such that each plot is given a numeral like 1,2,3 *etc.* In this way, if the orchard contains 12 cropping plots, these should be ascribed as Plot No... 1,2,3 and thus, upto the No. 12 to the last cropping plot.

Prefixing geographical direction to the number may be another system. In this system, both directions and numerals should be considered together and the plots that are located at the eastern, northern, western and southern sides might be assigned as East Block No., North Block No., West Block No. and South Block No. Ascription of cropping plots after the names of persons may be preferred by some to attach importance to some persons and they are be persons of eminence.

In whatever manner a cropping plot is recognized, the details of each plot should be recorded. For this purpose, a *separate register* should be maintained, in which particulars like area of the plot, its length, width *etc.* should be mentioned. Besides, the edaphic condition of the plot, *e.g.*, depth, texture, chemical reaction, analytical report if any and other specifications should be recorded. Suitable boards or pillars indicating assignment of the plots should also be installed at the boundary of each plot.

Chapter 14

Planting Distance

14.1 Principles and Objectives

Every organism occupies some space on Earth. Even a virus particle covers a space of 20-500 nanometer in diameter and 800-1200 nanometer in length. Very large space with a circular span of 14,500 square metres has been occupied by the banyan tree standing at the Acharyya Jagadish Chandra Bose Indian Botanic Gardens in Howrah, West Bengal. Among fruit plants, examples may be cited of say, pineapple which spreads about 90-150 cm area while chestnut tree that spreads to an area of 10-12 metres. If the fruit plants do not get adequate space, their growth and productiveness are greatly hampered.

In consideration to the above aspect, fruit plants are planted in the orchard such that they get adequate space to grow freely without obstruction yet the orchard space is not lost. That is, they are planted keeping in view their *canopy area* that would be assumed by them when they are fully grown up. This is referred to as planting distance of fruit plants. Thus, if a fruit plant assumes a canopy diameter of say, 10 metres in its fully grown up condition, theoretically it seems that its planting distance should be 10 metres. But from the practical point of view, some more space also becomes necessary to be provided to it. That is, if the canopy would assume a diameter of 10 metres, planting distance which is to be given should be little more than 10 metres, *i.e.*, 10.5 to 12 metres. However, if more than that distance is provided, there would be economic loss, *i.e.*, number of fruit plants to be accommodated in the orchard would be less. The distance which is measured between rows of fruit plants is regarded as the *row-to-row* distance while the distance between the fruit plants within the same row is regarded as *plant -to-plant* distance.

Planting distance has in fact, a great impact in fruit culture. If the distance is less than the normal, normal growth of the fruit plants and their roots get hindrance. Due to that, their productiveness is greatly hampered. *Air drainage* inside the fruit plants is also much hampered resulting less availability of oxygen to the leaves. Inter-cropping would have difficulty in close spacing of fruit plants. On the other hand, spacing more than normal would accommodate less number of fruit plants in the orchard. Very wide spacing may result in much growth of weeds as well.

14.2 Factors Affecting Planting Distance

How much planting distance should be given for fruit plants in establishment of a given orchard should be determined in consideration to a number of factors of which, the following are of pertinence.

1. Species of Fruit Plant

Botanical species of a fruit plant is by far, the most important factor that determines its diameter of canopy in its fully grown up condition. This in turn, is the determining factor for its planting distance.

It is evident that *diametric spread of canopy* of any plant is specific to the species. Thus, the canopy diameter of a mango tree is specific within a certain range and could not be considered same as with all fruit trees. It should be remembered that when the propagating materials of a fruit plant are planted in the orchard, they are small in size. But every year, they go on increasing the spread of the canopy as well as the height. In this way, the girthwise and lengthwise growth are continued in them upto a certain age depending on plant species and after attainment of that age, any large increase does not normally take place. Thus, although a fruit plant occupies lesser space in early years of its life, much larger area is actually covered by it when it is in fully grown up stage. This is largely dependent on the species to which it belongs.

The above phenomenon needs to be considered to determine the planting distance. For example, let us presume that the sidewise spread of the fruit trees of a species attains a diameter of about 10 metres when they are fully grown up. Now, if the young propagating materials of it are planted maintaining a planting distance of much less than 10 metres, say 7 or 8 metres, the branches and foliage parts of the trees will obviously touch and overlap with those of the neighbouring ones when they will be in grown up stage (Figure 14.1). Due to such entangled condition, the branches and shoots of the trees will not get ample space to grow freely sidewise, besides that the inner parts of the trees will not receive adequate sunlight and air. As a result, normal growth and vigour of the trees will be seriously hampered and the trees will assume lanky appearance. In such congested situation, there will also be the problem of ventilation in the inside of the trees and due to poor *air-drainage*, which is especially apparent in cold weather, the vitality of the trees will be greatly impaired.

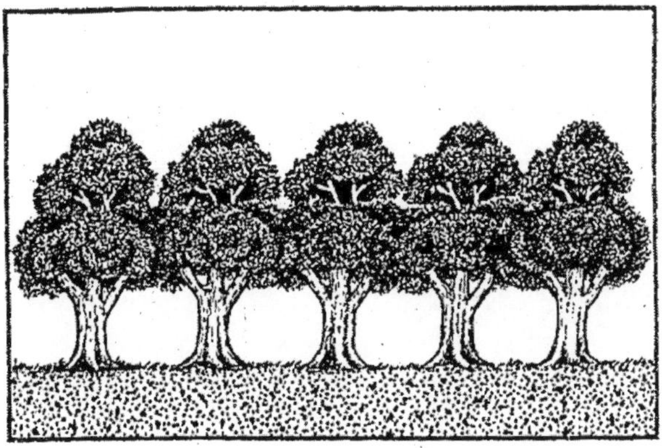

Figure 14.1: Branches and Foliage Parts of Trees have Overlapped with those of the Adjacent Ones due to Close Spacing.

On contrary to the above, if the propagating materials are set very widely apart, *i.e.*, say, maintaining a planting distance of 15 metres, their branches and leaves will get ample space to grow freely on all sides and will not overlap with those of the neighbouring ones. The inside parts of the trees will get the required sunlight and air and thus, due to favourable condition, the fruit bearing parts of the trees will also become higher. The ultimate result is, both yield as well as quality of the fruits will be high for each individual tree in the orchard. Hence, seemingly, one would consider it justifiable to maintain wide distances in planting the propagating materials in the orchard.

But as shown in Figure 14.2 due to such type of wide spacing, much space in between the maximum spread of the fruit trees is left open and remains unutilized. Accordingly, the total number of trees in the orchard will be on the lower side than when close planting is done. Thus, although per tree yield of fruits may be satisfactory in this case but it will not be so, if the total yield of fruits for the entire area of the orchard is considered as a whole. The reason is, lesser number of trees actually planted in the orchard area.

The two foregoing examples make it apparent that securing satisfactory, in other words, *optimum* yield of fruits from the orchard is not possible, if the trees are spaced either very closely or very widely apart. Besides this, quality of the fruits also becomes inferior in close planting as the foliage parts do not receive adequate sunlight and air. Since the basic objective of commercial fruit culture is to secure higher yield of fruits from the whole orchard and not only from one or few trees only and at the same time to harvest better quality fruits, he should not select such a planting distance which is very less or very wide. Instead of that, he should take up such a planting distance by which he is sure to get satisfactory yield of fruits and of better quality. This distance which is regarded as the optimum needs to be determined with due consideration to a number of factors among which the following are of significance.

Figure 14.2: Much Space in between the Maximum Spread of the Trees Remain Open and Unutilized Due to Wider Spacing.

It should be remembered that sidewise spread, that is, *crown of the tree* assumes more or less a circular appearance when it is in fully grown up condition. It has also been stated above that the girth of a tree increases every year upto a certain age of it and after that age, any large increase does not usually take place. The maximum lateral spread in mature condition of the tree is again, primarily dependent of the species. Such type of crown growth is apparent not only in fruit trees but also of fruits shrubs of fruits and the fruit herb, like banana.

Therefore, planting distance of fruit trees, shrubs or herb as the case may be should be determined in consideration to the species. (Although it is applicable to tree, shrub or herb, the term, tree has been retained in this text for convenience and as most of the fruit crops are in fact, trees.) Thus, for example, if the trees under a species assume a crown spread of 10 metres, the planting distance should also be 10 metres, by which there will be no overlapping of branches, nor there will be any wastage of land.

The above postulation though seems to get support from the theoretical point of view, but not so from the practical standpoint. This is because, some growths which may not be large are likely to appear at the periphery of the trees even after they have fully grown. Such growths arise usually after heavy pruning of the branches. In such case, the foliage parts would no doubt, act in overlapping with those of the next trees.

To get rid of this overlapping, some more space is therefore, needed to be left as allowance. That is, the correct planting distance should not only be in consistence with diametric spread of the trees but also with allowance of some space in addition to that. This allowance may be from 0.5 to 2 meteres but leaving a free space of 1 metre or so in between the lateral spread of canopy of the trees appears to be satisfactory in most cases. Thus, if the canopy diameter of the trees of a species goes upto 10 metres in full maturity, 11 metres should be maintained as the distance in planting the propagating materials.

As shown in Figure 14.3, by giving such planting distance, the leaves and branches of the trees will grow freely in an unobstructed manner and wastage of space in the orchard also does not happen.

Figure 14.3: Branches and Foliage of Trees have not Touched with those of Neighbouring Ones and Wastage of Space is Avoided.

2. Variety within Species

The height, growth habit and diametric spread of the canopy of a fruit tree are although determined by the botanical species, horticultural varieties, cultivars, types *etc.* existing within the species, these are also known to be important factors for growth characteristics in many cases, however, within certain limit. For example, in mango, the variety, Langra is observed to be more vigorous than the Dusheri and many other varieties growing under the same edaphic and agro-climatic conditions. The varieties like Mankurad, Pairi, Olour, Cowasji Patel, Baneshan, Banganpallee *etc.* are also some examples of a large number of vigorously growing varieties of mango. In case of banana, the Red banana variety is recognized as a tall, robust and highly vigourous one. In papaya, the varieties, Washington, Santacruz giant, Pusa giant *etc.* can be named as of high vigour, in contrast to the Pusa dwarf and Pusa nanha varieties which are dwarf and much less vigorous with lesser spread of crown.

Varietal difference in regard to the tree vigour is noted in many species of subtropical and temperate fruit trees as well. For example, in guava, the famouns varieties, Allahabad safeda and Lucknow-49, as also Dholka, Nasik and some varieties of Maharashtra exhibit high vigour than the varieties like Behat coconut, Hari Jha, Baruipur, Hafsi *etc.* while the shy bearing, Apple guava variety that is famed for producing apple looking fruits is much less vigorous. In grape, the varieties grown in the peninsular India such as Bhokri, Phakri, Kandhari, Kali sahebi, Bangalore purple, Anab-e-shahi, Pacchai drakhsai *etc.* are highly vigorous than the north Indian varieties like Khalili, Early Muscat, Perlette, Thompson seedless, Pusa seedless, Bedana, Delight *etc.* Some examples of the vigorously growing varieties of litchi are Seedless late, Rose scented, Bombai *etc.* and a less vigorous one is the variety, China. The variety, Flerte of avocado is one that is characterized by highly spreading growth habit.

Among the varieties of apple, Gala, Jupiter, McIntosh *etc.* are some of the many examples under the vigorously growing group and those like Jonathan, Jonagold, Jester *etc.* of the moderately vigorous growing group while the Greensleeves variety is one of the less vigorous group. Again, the varieties of pear, *viz.*, Anjou, Moonglow, Magness, Flemish beauty, Patharnakh *etc.* are more vigorous than those like Le Conte, Niiisseki *etc.* The virorous varieties of peach include Redglobe, Early amber, Dixigen *etc.* and in plums, Victoria, Bradley's king *etc.* are mentionable.

In determining the planting distance, the orchardist should therefore, take into consideration the variety of the species of fruit tree also which he is going to grow in his orchard. Evidently, the planting distance should be suited to the tree vigour of the variety, such that the more vigorous varieties are to be spaced at somewhat greater distance than the moderately vigorous ones while those which are less vigorous are to be set at still lesser distances.

3. Climatic Condition

Climate is one of the most important environmental factors that affects vigour of a tree. Among other climatic components, the rainfall and atmospheric temperature are however, known to exert greater influence on the tree vigour. Thus, high rainfall continuing over seasons along with mild winter offer congenial condition for increased vigour of the trees, resulting dense and extended lateral spread of their branches. The luxuriant and spreading growth of the fruit trees grown in such climatic condition obviously necessitates greater planting distance of them.

Under Indian condition, high vegetative growth of fruit trees is pronounced greater when they are grown in the tropical situation of south India and especially, in the coastal tropics. Accordingly, in setting up of a fruit plantation, the propagating materials are spaced wider in that zone than when they are grown in the milder tropics or in subtropical climate. For example, in the coastal areas of Tamil Nadu and Kerala, Basrai dwarf variety of banana is planted with a spacing of more than 2 metres but in the East Khandesh district of Maharashtra, where the climate is drier, plants of the same variety are grown with a spacing of 1.3-1.5 metres only.

The mango trees grown in the high rainfall area in the peninsular India assume dense, vigorous and highly spreading growth and hence, they are grown there maintaining a planting distance of more than 12 metres. Under north Indian condition on the other hand, the trees are not so spready and luxuriant in growth due to relatively low rainfall, distinct winter and summer, for which the fruit tree is grown with a spacing of 8-10 metres. Similarly, in heavy rainfall area in the south, the tropical fruit tree, spota takes up highly vigorous growth and accordingly, it is grown with a spacing of 9 metres or more. In the milder tropics of West Bengal and especially in the South 24- Paraganas district, sapota orchards are however, built up maintaining a planting distance of not more than 8 metres. Again, in the still drier parts of the Bhagalpur district and other eastern parts of Bihar, sapota trees are much less vigorous and thereby, necessitate a spacing of less than 7 metres.

In the Pune region of Maharashtra, both the purple and the green varieties of avocado are grown with a spacing of 8-9 metres. But in the coastal area of south India and also in the alluvial soil, wider spacing of upto 12 metres is given in conistent with the luxuriant growth of the trees.

Apart from rainfall, high wind is a notable climatic factor that is taken into consideration in determining the planting distance, especially for those fruit trees which are more vulnerable to wind. In such cases, close spacing is preferred by the fruit growers to protect the fruit trees against desiccating effect of wind. For example, in some parts of central India that are threatened by high wind, the mandarin orange is grown maintaining a planting distance of 5 metres or even less. Similarly, when litchi is grown in north India, less than 8 metre spacing is maintained to protect against the *looh* wind of summer.

4. Soil Condition

Edaphic factors, notably textural condition, depth and fertility of the soil greatly determine growth and vigour of the trees, their nutritional status and above all, fruiting performance of them. In accordance with, spacing of fruit trees is determined so that they become well-suited to the type of soil where they are grown.

The normal tendency is to provide relatively wide spacing when the fruit trees are grown in fertile land than when the same trees are grown in poor soil. Thus, sweet orange and mandarin orange trees are spaced at 6-8 metres when grown in fertile soil and this is reduced to 5-6 metres when grown in poor soil under the conditions of Maharashtra and in north India. In guava also, it is noted that in poor soils in the hilly regions of Maharashtra and south India, the trees are planted with a spacing of 5-6 metres while in the alluvial soil along the river banks of south India where the trees assume large appearance, they are planted with a spacing of 7-8 metres.

Similarly, pomegranate is an example which is grown in the poor soils of south India at a planting distance of 5 metres or less while in the fertile, alluvial soils of Gujarat and parts of north India, it is grown with a spacing of 7 metres or more. Jackfruit trees deserve attention in that, they are very much responsive to soil fertility and make prolific growth when grown in rich soil. Accordingly, in the highly fertile alluvial soils of south India and Odisha, the trees are given extremely wide spacing. Wide space is also provided to this fruit tree on the gangetic alluvial soil of West Bengal and Bihar. But in the poor lateritic soil in some parts of southern Jharkhand and in the western hilly area of West Bengal where jackfruit is abundantly grown not only for the ripe fruits but also for the immature fruits used for cooking, the seedlings are set in blocks with a distance of less than 9 metres.

Custard apple is a classic example which grows and fruits well in inferior type of soil and hence, it is cropped for utilization of waste land also. Normally, the fruit tree is grown with a spacing of 4-5 metres as is the case in the red sandy and gravelly soil of Andhra Pradesh known as *chalka* soil where it is grown to a large extent. However, the tree makes vigorous growth if its roots find adequate plant nutrients in the soil and thus, in rich and deposited soil in Jharkhand, seedlings are set in plantations at wide distance.

The above discussion therefore, points out that the planting distance should be lesser in poor soil than when grown in fertile land. However, it could be stated that the opposite viewpoint also gains support from many fruit growers as well as fruit scientists. An example may be cited as of guava in the eastern Uttar Pradesh, which is the most important guava growing area in India. Many guava growers in that belt are of the opinion that if the soil is not so fertile, it is better to crop the fruit tree with wider planting distance, in order that roots of the tree are able to forage over a long distance in search of food without facing competition with roots of the neighbouring ones. This viewpoint is held by many fruit growers in different parts of India, in respect of many a fruit crop.

Apart from fertility, other characteristics of soil also affect growth of fruit trees and thereby, the planting distances of them. In loose-textured soil with adequate organic matter, the roots of a fruit tree extend horizontally over a large area. As the spread of roots of a tree causes the *above-ground parts of it also to spread*, the fruit trees that grow in such type of soil become vigorous and spready. Accordingly, planting distance should be greater in this soil condition. In heavy soil on the other hand, the roots are compact, do not spread much in sides and in consistence, the planting distance should also be lesser in such type of soil.

The depth of the soil is another important factor in fruit culture, owing to the fact that *anchorage roots* in most of the fruit trees tend to move deep into the soil. Thereby, the growth of the trees is stunted in shallow soil and vigorous in deep soil. The planting distance should therefore, be adjusted accordingly.

5. Propagation Method

The method of propagation that has been adopted for the fruit tree is undoubtedly, a factor of tremendous importance in determining vigour of it and in turn, the planting distance. It is a common observation that a tree raised from seed, (precisely speaking, from the fertilized embryo of the seed) becomes more vigorous in growth than when it is propagated by vegetative means.

However, unless when it is necessary as in papaya, seed propagation is not advisable for fruit growing in commercial scale, because of the fact that the trees evolved from seeds fail to maintain true-to-type characters for high degree of heterozygosity in their chromosomal make-up, eventuated by high degree of cross-pollination. Nevertheless, seed propagation sometimes becomes necessary, especially when mass planting is done. Seed propagation of fruit crops is in fact, an age-old practice in India and as the trees produced from seeds are usually hardy and quantitatively more productive, many fruit growers in this country still prefer this to erect orchards in adverse type of soil. Evidently, the planting distance should also be made higher when these propagating materials are to be used.

Among all vegetative propagation methods, it is *grafting*[4] which influences the vigour of the tree and this is due to the effect exerted by the *rootstock*. As a matter of fact, in many species of fruit trees, a number of suitable rootstocks have been produced which make the fruit trees to grow either as dwarf, compact and less spready, or, as tall, vigorous and more spready when the latter are used as *scion* materials. Few examples of these rootstocks in different species of fruit trees have been cited in the following.

Maximum attention has perhaps been paid in apple, aiming at evolving such rootstocks as would suitably modify the growth of the scion trees of apple of different varieties. The East Malling Research Station in Kent, Maidstone, England deserves credit for production of rootstocks of this type and these are universally recognized as Malling rootstocks, abbreviated by the letter, M and bearing numbers in Arabic. The Malling rootstocks have though great influence on vigour of the scion trees, they have some drawbacks and especially in having susceptibility to the insect pest, wooly aphid. To get rid of this problem, the John Innes Horticultural Institute in Merton, England and the East Malling Research Station have jointly prosecuted breeding work in apple with a view to evolve resistant rootstocks. The rootstocks produced in this scheme are also designated numerically but prefixing with the letters MM, denoting thereby, Malling-Merton.

Among the series of rootstocks released by the said Institutes, those which make the scion trees to grow more vigorously are M-12, M-16, M-25, MM-104 *etc.*, those which make the scion trees to become dwarf and much less vigorous are M-9, M-20, M-27 *etc.*, while those which make the scion trees to assume semi-dwarf appearance include M-2, M-7, M-26 *etc.* It needs pointing out that among all these rootstocks, M-9 has *most dwarfing effect* while M-16 has *most virorous effect* on the scion tree.

In pear, a widely used rootstock that produces dwarfing effect on the scion tree is quince, which belongs to a different species (*Cydonia oblonga*). Some dwarfing rootstocks in pear have also been clonally produced in South Africa.

The rootstocks in plums which impart greater vigour to the scion trees are Black plum, Marianna-4001, Myrobalan-2-7, 29 C *etc.*, while less vigour is produced by Marianna-2624. In apricot, MP-43 and especially plum rootstock are known to produce dwarfing effect to the scion.

Rootstock effect on scion has been extensively worked out in citruses also. The trifoliate orange (*Poncirus trifoliata*) which does not belong to the genus, *Citrus* convincingly exerts dwarfing effect when used as a rootstock for citrus trees as scion. materials. Other rootstocks, notably, rough lemon (*Citrus jambhiri*), karna khatta (*C. karna*), Rangpur lime (*C. limonia*), sweet orange variety, Mosambi *etc.*

4. Grafting refers to both scion-grafting (syn. Grafting) and bud-grafting (syn. Budding).

impart increased vigour to the scions while the varieties of citrange, mandarin orange *etc.* are recognized as semi-dwarfing rootstocks for different *Citrus* species.

For guava, the related species, *Psidium pumilum* has been shown to exert dwarfing effect on the Allahabad sareda variety used as scion. By the use of quince rootstocks, loquat trees are known to assume dwarfness. Increased vigour of sapota trees is also imparted when *khirnee* is used as a rootstock of it. Some varieties or perhaps type of seedlings of ber also prove to be dwarfing rootstocks and these are used for propagating ber in some parts of Gujarat and north India.

The above discussion thus, makes it apparent that rootstocks have a profound influence on the vigour of the scion materials in case of many fruit trees. Hence, the planting distance needs also to be adjusted in accordance with, such that greater distance is maintained for virorous rootstocks and lesser distance when dwarfing and less vigorous rootstocks are used for propagation.

6. System of Layout

The distance to be maintained in planting the propagating materials should also depend on the system of layout that is actually followed in the orchard. As stated in Chapter 12, there have been a number of systems in laying out the cropping plots in orchards and evidently, for any particular species or variety of fruit tree, the planting distance will not be the same under all systems.

Thus, in the rectangular system of layout, the row to row distance remains much greater than what is followed in the square system of layout. Again, the row distance is somewhat lesser in the triangular system of layout than the square system. Both the row to row and tree to tree distances remain however, lesser in the hexagonal system of orchard layout as compared to the square system. In the hedge-row system which is practised for some particular purpose, the propagating materials are set loosely apart in the row and wide distance is maintained between the rows. In the contour system of layout which is followed in hilly area, the distance between trees and rows may in fact, be changed in the same plantation whenever it is necessary in accordance with the acclivity or declivity of the terrain.

Even the runner producing fruit plant, *i.e.*, strawberry which grows and fruits on the ground, planting distance of runners depends on the planting systems. For example, in the matted-row system of planting, the plant distance remains 0.35-0.50 metre, in rows of 0.7-0.9 metre, in spaced-bed system, runners are planted 0.5 metre apart in rows spaced at 1 metre while in the hill-row system, planting is done in single or double rows at a spacing of 0.25-0.35 metre.

7. Manures and Fertilizers

Addition of manure improves the texture, water-holding capacity *etc.* of the soil besides adding some nutrients and thereby, enables the roots to ramify and grow freely. The application of fertilizers enriches the nutrient status of the soil such that the roots are not suffered for food. Therefore, application of these inputs of suitable types, in suitable proportions and time make the trees to grow vigorously and necessitate wider spacing.

8. Irrigation

While scarcity of water stunts growth of the fruit trees, assured supply of irrigation makes them vigorously growing. The system of irrigating an orchard is however, a contributing factor for growth of the fruit trees. For example, when irrigation is applied by flooding,exuberant growth of weeds becomes a common feature in the orchard. As a result, although the roots of the fruit trees are supplied with copious amount of water, they are to face competition with the weeds for nutrients, which stand in the way for proper growth of the above-ground parts. The problem does not appear to be so serious in ring or basin system of irrigation, where the tree makes good growth, both lengthwise and girthwise. Higher growth of trees thus, necessitates higher planting distance of them.

In *drip irrigation*, root hairs of the fruit trees get timely supply of water and besides that, high population of weeds in the orchard land does not become a problem. Thus, the fruit trees would seem to make highly vigorous growth in an unoostructed manner. As a matter of fact, the vegetative growth of the trees is though optimum in drip irrigation but not very high. The sprinkler irrigation on the other hand, is noted to make more spready growth of the trees in most cases.

However, it should not be forgotten that drainage is a factor which is of no less importance than irrigation in fruit culture. Thus, whatever efficient the irrigation is, growth of the trees will be seriously hampered if the drainage is poor.

9. Soil Management Practice

The growth of fruit trees is also much influenced by the soil management system that is practised in the orchard. Each of the different systems has its own merit, if intelligently followed and at the right situation and if not, may prove to be even much harmful.

In *clean culture* system, the land is ploughed and kept free from weeds. For this, the roots are not to suffer in competing with those of the weeds for moisture and nutrients. Further, soil aeration is good and percolation of water in the soil is also not impeded due to hard crust. As a result, the trees make good vegetative growth. On the other hand, many roots of the trees are also cut off or damaged due to ploughing done in this system. This is more common in shallow rooted fruit trees like papaya, guava, sapota, litchi *etc.* and hence, their growth is seriously hampered. Erosional loss of soil may also be high in this system resulting tilting of the trees or exposure of the roots, if the land is not flat.

In *sod culture* system, the grasses or forage crops that are grown bind the soil in a compact manner for which erosional loss is overcome. Besides that, surface run-off after rain is less but due to uptake of much amount of moisture and nutrients by the grasses, the fruit trees are suffered. However, mowing down the grasses is a better practice which acts as an organic mulch and thereby, the trees make good growth. Growing *cover crops* is though done to conserve orchard soil against erosional loss, it is very much helpful for growth of the fruit trees when leguminous plants are grown that add nitrogen to the soil. Similarly, *green manuring* if done in the orchard adds organic matter and nitrogen, ensuring increased vegetative growth of the fruit tree.

Mulching is undoubtedly a highly effective soil management system for a number of reasons, especially in conserving soil moisture and disallowing growth of weeds. Among fruit trees, those which are shallow-rooted are especially benefited by mulching and papaya is a notable example, which makes vigorous growth when the soil is covered with a suitable mulch.

Evidently, in determining planting distance for the fruit trees to be grown, the orchardist should bear in mind what soil management practice is to be done by him in running condition of the orchard.

10. Filler Tree and Inter-crop

When the orchard is run on commercial basis, the growers usually become interested to earn an extra profit by raising some supplementary crops. This is done by growing *filler* trees, or *inter-crops*, or both.

The filler trees or plants comprise short duration fruit trees or plants, *e.g.*, papaya, banana, phalsa *etc.* which are planted in the diagonal point of each square and are cropped temporarily. Many orchardists presume that the planting distance maintained for the main fruit trees that are planted in each corner of a square would remain the same whether the middle part of each square is utilized by growing filler trees or not. But this is not correct in most cases, because this may result in a crowding of the main trees and the filler trees. Hence, if at all, filler trees, i.e., supplementary fruit trees have been decided to crop in the mid-part of each square, planting distance for the main trees, *i.e.*, those which are to be set at each corner of a square also needs to be widened to avoid crowding and competition between these two types of fruit crops.

When inter-crops (*vide* Chapter 19) are grown, the fruit growers usually have the tendency to crop them in blocks. That is, rectangular and wide strip of land is left between rows and the rows may be 1 to 5 or more in number. If such type of cropping is done, evidently to make provision of the strip of land for raising the inter-crops, the row distance also needs to be widened accordingly.

11. Training and Pruning

A particular size, shape or form is imparted to a fruit tree by the operation of *training*, that is initiated in the first few years of its growth. For erect growing fruit trees, three systems of training are in vogue, which are (i) central leader, (ii) open centre and (iii) the modified central leader systems as are shown in Figure 14.4 (a), (b) and (c) respectively.

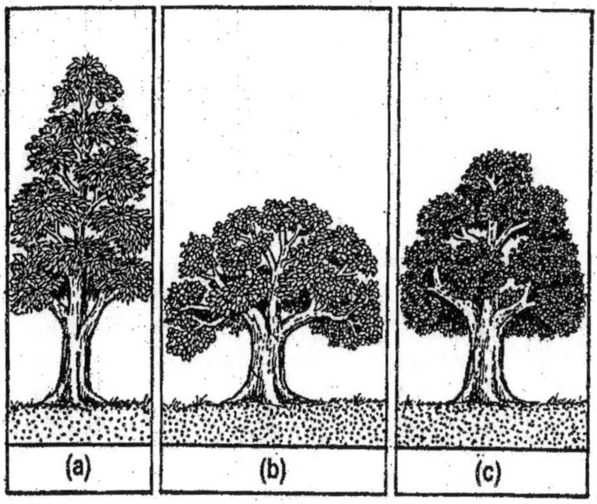

Figure 14.4: Training of Fruit Trees by (a) Central Leader, (b) Open Centre and (c) Modified Central Leader Systems.

In open centre system, the tree assumes a tall appearance and lateral spread is less while the open centre or vase system is just the reverse in which case, the lateral spread of branches is much more with lower height of the tree. The modified central leader system is an intermediate between the above two that makes a tree neither very tall nor highly spready in girth. Accordingly, the planting distance to be given for the fruit trees should be lesser, wider and medium, if it is decided to train them by the central leader, open centre and modified leader systems respectively.

Specialized systems of training are however, followed in viticulture to provide support to the grapevines. In this fruit crop, for less vigorous varieties commonly grown in the Punjab and Haryana, lesser planting distance is provided when trained by the head, vase or goblet system which does not need extra support and somewhat greater spacing is given in the telephone system. Spacing remains greater for other systems followed in south and west India, *e.g.*, single stake, cane, espalier, bower, cordon *etc.* while much greater planting distance is necessary to be given in overhead trellising (= Pandal)system erected for growing highly vigorous varieties in many parts of south India.

The act of pruning, *i.e.*, cutting off any part of a fruit tree which affects function and thereby, its productivity also trains it, though indirectly. In thinning-out system of pruning, the shoots and branches which are to be pruned are cut off completely at their bases leaving no stub and those which are not pruned are left as such to grow them naturally. Accordingly, the branches grow in length and the lateral spread of the tree becomes higher. The reverse system is heading-back, in which case, the shoots and branches that are to be pruned are cut off at their upper parts only while the basal parts are not disturbed, in order to encourage emergence of new shoots below the cut surfaces. Lateral spread of the tree is thus, relatively less in this system than the above. Again, some trees are drastically pruned of their branches so as to maintain less spreading shape of them. In determining the planting distance, what type of pruning should be done for the fruit trees which are to be grown in te orchard is also therefore, a consideration.

12. Lodging

For shallow-rooted fruit crops in particular, lodging appears to be a serious threat in many parts of India, the causes of which may be wind, washing away of the surface soil, intercultural operations, attack by wild animals *etc.* The trees may be lop-sided and lie aslant or are even uprooted and lay flat. The pineapple plant is very much vulnerable to this hazard as its roots are scanty, not well-developed and these abound only on the surface soil. The problem is more common in the pineries in Tripura, Assam, Meghalaya and other eastern States and in the Jalpaiguri district of West Bengal than when the fruit crop is grown in south India. Papaya can be ranked as the next one to suffer from this calamity and other examples which are to be included are litchi, sapota, banana *etc.* Close spacing is a practical solution in such cases that ensures mechanical support to the plants by the adjacent ones.

13. Improving pollination

At least in custard apple, *close planting gives rise to convincing result in improving pollination*. The fruit tree though produces flowers from summer to rainy season but those borne in summer very often fail to set fruits due to low relative humidity. The condition is apparent in Egypt, Israel, other middle-east countries and in India, in the drier parts of the north Bihar, southern Jharkhand and the adjoining Purulia and Medinipur districts of West Bengal. Close planting greatly averts this problem by ensuring increased relative humidity in the outer and inner parts of the trees. Setting the trees with a close spacing of 2.5-3.5 metres has been found to give encouraging results than the normal spacing of 4-5 metres in the Purulia district of West Bengal where rainfall is low.

14. Protection against Diseases

Close planting of fruit trees sometimes becomes necessary to protect them against diseases and at least in papaya, this is found to be much effective. It is often seen that the Coimbatore varieties of papaya become highly infected with diseases when grown in drier climate. This usually does not take place or the intensity of disease incidence is reduced by planting the seedlings at closed or even at very close distance, where the relative humidity of the *micro-climate* is increased. High relative humidity though appears to favour disease incidence but the effect is opposite as is observable in a mist propagating chamber. (where very high humidity is conducive to root formation).

15. Dwarfing by Plant Growth Regulating Chemicals

Dwarf fruit trees are sometime preferred by many orchardists as training, pruning and harvesting operations become easier and overall management becomes more efficient and less costly. Although cropping of less vigorous and dwarf fruit trees is normally done by the use of suitable varieties, rootstocks or training methods, nowadays, many fruit growers feel interested to induce this by the use of chemicals. Plant hormonal chemicals, notably those recognized as *plant growth retardants* are most effective in this respect which retard growth at the *sub-apical meristem* of a plant. Beyond doubt, these have to be applied when the fruit plants are in early stage of growth.

A host of such growth retarding chemicals are available, the trade names of which are AMO 1618, AMO 1619, Phosfon-D, Cycocel, B-995, CO-11, BOH, Morphactin, Nicotinium *etc.* However, effectiveness of them is greatly dependent on the selection of the chemical, concentration used and the stage of growth of the plant. Evidently, planting distance should be lesser for the culture of dwarf and less vigorous fruit plants. The compound, Chlormequat is much used now for its high effectiveness.

14.3 Normal Planting Distance of some Fruit Plants

The above discussion makes it overt that planting distance of fruit crops is influenced by a number of factors, among which some have major role. Nevertheless, a list has been provided below in table 14.1 which has presented the spacing that would be required by fruit crops belonging to different botanical species, if grown singly in the orchard or in a plantation. It should be noted that the list is only a general one and a range of planting distance has in fact, been provided for each fruit crop. Hence, this should not be considered to be applicable under all conditions. Although the list may be considered, the actual planting distance should in fact, be determined with close attention to the factors associated with it.

Table 14.1: Planting Distance of some Fruit Crops

Fruit Crop	*Range of Planting Distance (Metre)*	*Fruit Crop*	*Range of Planting Distance (Metre)*
Almond	5-6	Litchi	8-10
Apple	8-9	Longan	7-9
Apricot	6-8	Loquat	6-8
Avacado	7-10	Mandarin orange	6-7
Bael	8-10	Mango	9-12
Banana	2-3	Mangosteen	8-10
Ber	8-11	Mulberry	2-4
Cape-gooseberry	0.5-1.0	Olive	6-8
Carambola	6-8	Papaya	2-4
Cashew	7-9	Peach	5-7
Cherry	8-10	Pear	4-6
Chestnut	12-14	Pecan	15-17
Custard apple	4-5	Phalsa	3-4

Fruit Crop	*Range of Planting Distance (Metre)*	*Fruit Crop*	*Range of Planting Distance (Metre)*
Date	4-6	Persimmon	7-9
Dillenia	7-10	Plum	5-6
Elephant apple	8-9	Pomegranate	3-6
Fig	7-9	Pistachio	5-7
Filbert	5-7	Rose apple	6-8
Grape	2-4	Sapota	8-10
Grapefruit	5-7	Shaddock	7-9
Guava	4-7	Sour lime	4-6
Hogplum	5-8	Star apple	7-9
Indian hog plum	9-12	Strawberry	0.3-0.5
Jackfruit	9-12	Sweet lime	6-8
Jamun	10-12	Sweet orange	7-8
Kiwi fruit	4-6	Walnut	8-10
Lemon	7-8	West Indian/ Barbados Cherry	4-6

Chapter 15

High Density Planting

15.1 Definition and Concept

An orchardist has the firm belief that in efficient farming of fruit crops, the distance between the rows of fruit plants and between the fruit plants in a row should be properly maintained. This is with the idea that if the distance is less than optimum, there will be overlapping of shoots between the plants which will effectuate a competition among the shoots and the shoots will be shaded. Also that, there will be competition among the roots of the adjoining fruit plants. Owing to such undesirable or anomalous situation, *i.e.*, deviation from the normal system of planting, it is estimated that the growth and productiveness of the fruit plants as well as quality of the fruits will be seriously hampered. On the other hand, if the distance between the rows or between the fruit plants in the rows is kept more than optimum, there will be much wastage of land in the orchard.

In accordance with the above concept, optimum space is maintained between the rows and between the plants in the rows. How much space should be considered optimum however, varies according to the species and variety of fruit plant, soil and climatic condition, cultural practices to be adopted and other factors. These have been elaborated in the Chapter on Planting Distance (Chapter 13).

On contrary, *high density planting of fruit culture* is an innovative practice in which, the distance between rows and between fruit plants in the rows are kept low or very low. Hence, this system could be said to be a challenge in the conventional way of fruit growing and the trend goes by the term as *high density planting,* or *close-density planting*.

As a matter of fact, high density planting is grouped under three types. These are termed as (*i*) *high density or, medium high density planting,* (*ii*) *optimum high density planting and* (*iii*) *maximum or, ultra-high density planting.*

In the medium high density planting, fruit trees are trained to impart bushy appearance. Those fruit trees which are not sustainable to very close spacing are preferred for growing by this system. In general, upto 2000 trees are accommodated in a hectare area of land in this system. The number of trees are made increased upto about 12,000 in the optimum high density planting system. On the other hand, in case of ultra-high density planting, extremely close spacing is given and thus, even upto a lakh of trees are planted in the same area of land.

Apparently, it seems that high density planting could not be said to be a scientific way of fruit culture. But the practice which was first established in the nineteen sixties in Europe in the culture of apple trees was found to be very successful in enhancing productiveness and the quality of fruits. Thus, high density planting of fruit plants is a novel innovation in fruit culture.

15.2 Principles

In high density planting, number of fruit plants is a specilized system which per unit area of orchard land is higher or much higher than those of the conventional type of orchards where spacing between fruit plants remains to be higher, *i.e.*, grown with normal spacing. Hence, number of fruit plants being higher, yield per unit area of land in a high density orchard goes high. As a matter of fact, the fruit plants that are grown in a high density orchard are very low-heighted and that, their branches make very less sidewise spread. Accordingly, the number of fruits that are borne in a fruit plant which is grown in a high density planted orchard is *far less* than a fruit plant which is grown in an orchard in which normal spacing of fruit plants has been given. Thus, number of fruits in a fruit plant grown in a high density orchard is much on the lower side. But as the number of fruit plants in a high density orchard is very high, the *total yield per unit area* goes very high.

In fact, high density planting of fruit plants is a specialized system which involves *high intensification* in the culture of the fruit plants. It is because of the the intensive culture and management, high yield is obtained in orchards of such type. This becomes possible as all parts of each fruit plant in a high density orchard are within easy reach of the orchardist. For this advantage, management could be done in a much better way. Fruit plants in a high density orchard are cultured somewhat like field crops and hence, many people jokingly call it, "agronomic way of fruit farming".

15.3 Special Features

In establishment and management of high density planted orchards, attention should particularly be given on certain aspects. These are as follows.

Dwarfness of fruit plants is most important consideration. Hence, dwarf varieties, types, strains or hybrids of fruit plants are preferred. Fruit plants which have been grafted or budded on such rootstocks that make the scions dwarf or semi-dwarf are also preferred.

It may be mentioned in this connection that *plant growth retardants* (as stated earlier)which comprise a group of plant growth regulating chemicals tend to retard upward growth of the main shoot and branches of a plant and thereby, bring about dwarfness in it. Unlike plant growth inhibitors, which act on apical meristems, the plant growth retardants act on sub-apical meristems and *slows down cell division* and *cell elongation of shoots* without any formative effect. Apart from inducing dwarfness, the plant growth retardants have other beneficial effects also, such as strengthening of stem, increasing stress resistance, increasing chlorophyll content of leaves *etc.* As mentioned in Chapter 14 (Ref. 14.2-15), a large group of plant growth retarding chemicals are now available and among these, Chloromequat is much used.

To restrict upward growth of the fruit plants, proper *training* of them is by far, of great value. Among the different systems of training, the *central leader system* is inadvisable to adopt. Training operation should be done to aim at encouraging high number of fruit bearing branches and minimum number of branches that give rise to structural framework. Branches are encouraged such that their leaves do not provide much shade on leaves of other branches.

In adoption of the training method and staking of the trees, cultural practice followed, method of harvesting and other factors, several sub-groupings are made. For example, in the *tatura* trellising system developed in Australia, only two branches with suitable crotch angle are maintained in each tree that grows east and west and the rows run in north-south direction. In the *pyramid* system, the tree assumes a pyramidal appearance because of the central leader system of training that is followed and this is known to be adaptable in soil of low fertility. The *cordon* system refers to training the tree like a cardon. The *mechanized* system facilitates harvesting by mechanical means such as with a combine harvester and in *meadow* system as the term implies, harvesting is done by moving round the orchard.

Pruning is a must in high density orchards. Not only annual pruning in case of temperate fruit plants but it also becomes necessary for subtropical and even for tropical fruit plants in some cases. In some years, even drastic pruning may become necessary. For some fruit plants, root pruning at interval of few years is found to give high fruiting performance.

15.4 Advantages

High density planting is becoming increasingly popular in fruit culture in many countries particularly for the following special advantages.

(i) *High productivity* of fruits per unit area of orchard land as compared to orchards where conventional planting distance is given is the most important attribute in high density planting of orcharding. Evidently, this is due to growing much higher number of fruit plants in high density planted orchards, i.e., increasing polulation of fruit plants.

(ii) Along with higher productivity of fruits, *quality* of fruits borne by the trees in high density orchards is also usually higher than in traditional orchards where fruit plants are grown with normal spacing. Receiving optimum sunshine and air by the fruits in high density orchards is a possible cause for this benefit.

(iii) Precocity of bearing of the fruit plants grown in a high density orchard is a great merit. Their juvenile period being very short, gives rise to production in short time.

(iv) High density planting system of orcharding is *labour intensive* than in traditional orchards.

(v) Fruit plants in a high density orchard are easily manageable. Hence, their *interculture, training* and *pruning* operations could be done in efficient manner. Harvesting of fruits could also be done easily as the fruits are within easy reach of hand.

(vi) *Low cost* is involved in setting up high density orchards in most cases.

(vii) In high density pineapple plantations, apart from higher yield performance, the plants also produce higher number of suckers and slips.

(viii) Recurring cost of high density planted orchards is low.

15.5 Disadvantages

Despite many advantages, high density planted orchards have many disadvantages also and among these, the following could be brought to notice.

(i) The system is not suitable under all conditions. It cannot be adopted in land which is very *steep, shallow and unfertile.*

(ii) *Drought prone* areas are not suitable for setting up fruit plants with low or very low planting distance.

(iii) Cost of planting may be high. Plants require *staking* in many cases, even permanently and thus, involves high cost.

(iv) In high density orchards, vigour of some fruit plants is found to be low after some years.

(v) *Regular drainage* from the land is a very important consideration in high density orchards as water stagnates easily in such orchards.

(vi) It is an admitted fact that high density planting of fruit culture has not given encouraging response in the culture of some fruit crops in some parts of India.

(vii) Recurring expenditure is high in many cases.

15.6 General Opinion

High density system of planting of fruit plants proves its worth only when appropriate technology is applied. Evidently, the techniques are highly sophisticated and should be experimentally verified. It needs mentioning in this connection that unless adequate experiments are done to determine precisely the techniques that contribute to encouraging results, it is difficult to recommend it for large-scale application of this system under Indian condition. Experiments are therefore, needed to be done for different fruit crops in different agro-climatic zones in this country to work out the actual techniques for achieving real success in this system of planting. This is of great importance for recommendation of this system of orcharding in large scale under Indian condition, where it is an age-old practice to resort to traditional orcahrding.

15.7 Some Examples on High Density Planting of Fruit Crops

General effects of high density planting in some cammon fruit crops have been stated in the following:

- ☆ *Apple:* Few rootstocks are actually suitable for the respective scions to build up high density planted orchards in apple. Some of the dwarfing rootstocks, *e.g.*, M.9 M.4, M.7, M.26, Bud.9, J- TE-E, M.27, MM 106 *etc.* are however found to be promising. Central leader system of training is considered favourable and other systems, *viz.*, Slender spindle, Vertical axis, Hybrid tree cone *etc.* are also adopted. Tree height should not be more than 1.5-2 times of the clean alley. Leaf area index should not be greater than 1.00 over cropping zone. Higher yield is obtained with 70 percent light interception in high density apple orchards.
- ☆ *Pear*: Recent practice of high density planting in pear is hedge-row system, *i.e.*, when a uniform hedge is maintained in the row. Much less row spacing is also maintained in the system.
- ☆ *Peach*: Plants are set 3 metres apart in high density planting. In the systems known as *tatura and meadow*, planting is done at 5 and 1 metre distance and 2 and 1 metre distance respectively.
- ☆ *Plum*: Dwarfing rootstocks are used for high density planting and it has been reported that 6666 trees per hectare could even be planted.
- ☆ *Apricot*: In the Canino cultivar, 7200 plants per hectare has been reported for planting to build up high density orchards.
- ☆ *Banana*: Trials done in India revealed that yield had been high in high density planting but growth of plant had been slow. In other trial, it has been observed that about 30 percent of the plants could not be harvested in due time by adopting high density planting.
- ☆ *Mango:* High yield is found to give rise in mango by high density planting. In Coimbatore, mango trees raised by medium density planting with a spacing of 4.5 metres has been stated to give remarkable result and shortening of gestation (= juvenile)period of the trees by 3-4 years. Under north Indian condition, encouraging result has been observed in the Amrapali variety of mango when the trees have been set at 2.5 metre apart between plants and rows.

- A report has stated that to secure desirable result in mango by ultra-high density planting, the trunks of the trees should at first be grown upto attaiment of a height of 1 metre. Then the trees should be de-topped at 80 cm height. From the upper part of each de-topped tree, 3-4 side shoots that radiate on all directions should be carefully maintained. When leaves of the trees have become mature, 3-4 side-shoots radiating on all sides should be maintained in each of these secondary branches. The same process should be continued for two seasons. Fruiting should be allowed in the trees when their canopy area becomes 1 metre on all sides.
- *Papaya*: High density planting is found to give encouraging result in the dwarf papaya varities, such as in the cv. Pusa Nanha which is a very dwarf cultivar by raising the plants with a spacing of 1.2 metre both ways,
- *Guava*: High density planting has been found to give higher yield per unit area. But total sugars, soluble solids and the vitamin C content of the fruits borne by closely planted trees usually tend to decrease.(The actual techniques of high density planting in guava however, needs to be explored by experimentation).

Chapter 16

Procurement and Pretreatment of Propagating Materials

16.1 Source of Procurement

It should be well-remembered that an orchard is set up as a *long-term plantation*, in other words, it will have its existence in the same site for many years. Hence, if it is correctly set up, it will give desired result. But if any error is committed, adverse effect due to that fault will remain there and the adverse situation will exist as long as the orchard will exist there. Although little improvement to get rid of the fault could be done during running condition of the orchard but any drastic improvement becomes difficult to be made.

We know that an orchard is built up from the propagating materials. Accordingly, maximum attention needs to be paid on these materials. They should be true-to-type, genuine, as per specifications, properly made, free from undesirable traits, of proper maturity and highly efficient. Only such materials give rise to ideal fruit trees in course of time and result in fruits of higher yield and of better quality.

Nurseries are the source of procurement of the propagating materials. The materials should be purchased from such type of nursery(ies) which are known to be reliable and provide genuine materials. If the orchardist has no knowledge on nurseries he should consult with other orchardists or experts. The orchardist should confirm that the nursery from which materials are to be purchased should have government registration. A nursery which is near at hand should not be given the only criterion in procurement of materials. To secure the desired and the genuine propagating material of a plant species, the orchardist may have to go to a long distance. An important point is, it would be unwise for the orchardist to obtain inferior type of propagating materials at low price than to get the genuine type of material paying even higher amount for the same. He should not have "penny wise and pound foolish" attitude.

16.2 Quality of Propagating Materials

Before planting in the orchard, it should be confirmed that the propagating materials are of *desired standard*. These should be scrupulously examined considering the different aspects. Some of the criteria which need particular attention in judging the quality of the propagating materials are stated up in the following.

1. Presence of Adequate Roots

A propagating material is supplied by a nursery, maintaining the roots in an earthen ball, wrapped with polythene, or leaves or straw *etc.* or in a pot, as shown in Figure 16.1.

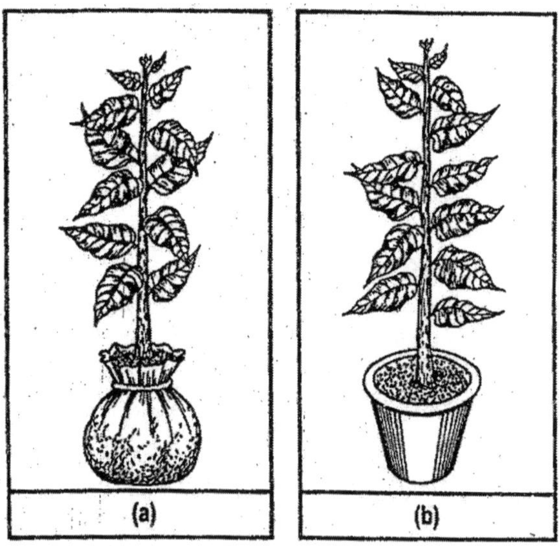

Figure 16.1: Propagating Material with Roots in (a) Earth Ball and in (b) Small Pot.

It is sometimes noticed that the earthen ball or the pot is small, *i.e.*, not proportionate to the part of the plant above the roots. Thereby, the roots which are the organs of the plant for taking up water and nutrients from the soil are not adequate in accordance with the shoot. This is usually done by the nursery to lighten the weight of the material so that transport becomes easier. The nurserymen frequently sprinkle water over the materials and maintain those in shade while selling. Therefore, the materials look fresh and healthy and the customers are allured.

The orchardist should not purchase such materials, because after planting, survival rate of such plants is very low for imbalance between uptake of water by the roots and loss of water from the above-ground parts. In other words, only such a planting material should be procured which has adequate roots, keeping in mind that higher the number of roots, greater is the chance of survival after planting it.

2. Adequate Callus Tissue in Rootless Stem-Cuttings

Propagating materials comprising stem-cuttings but without prominent exuberance of adventitious roots should not be used. Instead of that, cuttings, *i.e.*, rooted cuttings planted in pots should be preferred. Nevertheless, hardwood stem-cuttings are sometimes needed to be used particularly when the materials are brought from distant place. If these are used, it should be confirmed that adequate callus tissue has formed at the basal end of each cutting, axillary buds are bold in it and the cuttings have been dispatched covering in suitable water-holding material that contains moisture.

3. Insect and Disease Free Materials

There is no hesitation to the fact that for setting up of an orchard, only such propagating materials should be used that are healthy, in other words, free from the attack of any insect or animal pest or disease. Therefore, a propagating material should be scrupulously checked in all over its parts and both on the upper and lower surfaces of its leaves. In no case, it should be obtained if it shows any symptom that indicates the attack by a pest or a disease.

In this connection, it however, needs pointing out that although a plant may apparently look healthy, i.e., devoid of any disease symptom, presence of disease organism in its anatomical parts cannot be ruled out. Such cases are found for many viral diseases in particular, where the virus remains present in the body of the plant but does not produce symptom then. The symptom is in fact, exhibited later when the plant has grown much. From apparent and visual inspection, it is not possible to detect the existence of the virus and to do so, *indicator plants* are used. If the indicator plant produces symptom on grafting with the material, the existence of virus in the latter is indicated due to graft-transmissibility of the virus. Such procedure is not practically possible for each and every propagating material, besides that this will damage the materials. In the recent times, some nurseries and especially micro-propagation units are selling such planting materials, which are *certified* to the effect that they are free from any disease. These certified materials obviously cost higher but are safe to use. In *Citrus* species, certified materials or bud-sticks having no incidence of virus for the disease, *viz.,* quick-decline (*tristeza*), cachexia (*xyloporosis*), scaly bark (*psorosis*), scaly butt (*exocortis*) etc. are available from some sources. For grape and other fruit crops also, such materials are obtainable.

4. Free from Nutritional Deficiency

It should not be forgotten that health of a plant does not mean its freedom from the attack of pests and diseases only but also the absence of any nutritional deficiency. When the propagating materials are maintained in the nursery, they actually need all the macro- and micro-nutrients in required amount, which comprise their food. If these nutrients are available to them in optimum amount in the nursery, the metabolic activities in them take place in proper manner and thus, they make good growth. On the other hand, in absence of these nutrients, the plants are regarded as nutritionally deprived. Such a plant which is nutritionally deficient cannot grow in desired manner and therefore, should not be procured for planting in the orchard.

When a plant acutely suffers from any particular nutrient, it shows a characteristic symptom especially on its leaves, which is apparently recognizable. The symptom(s) of deficiency however, differs according to the nutrient, plant species and varieties. Nevertheless, the common symptoms that are produced due to deficiency of different macro- and micro-nutrients may be identifiable as *chlorosis* of leaves, which may be total, partial, veinal, intervienal, *etc.*, *bronzing* or *blackening* discolouration of leaves, *rolling* of leaves, *scorching* of leaves at the tips or the margins, *small size* of leaves, *misshaping* of leaves, *resetting* of internodes, *drying* of shoot tips, wilting of branches, *stunted growth* of plants and the like. The orchardist should closely examine a propagating material and be confirmed that such undesirable symptoms are not present on it.

5. Vigour and Branching

As a less vigours propagating material may not survive after planting, or does not make good growth, only vigorous material should be used. *Thickness of stem* is an important criterion as thin-stemmed material is regarded as of less vigour. Hence, thickness of the stem at different heights of it should be critically examined while procuring a material. For tropical and subtropical plants, number of leaves is also an indication of growth. Where branches are present, is should also be ensured that they are well-spaced and radiated on all sides. The age of the propagating material is by far, one of the most important considerations to select it for planting. But this actually depends on a number of factors. For example, if seedlings are used as in case of papaya, one to three month old seedlings may be transplanted. If cuttings (= rooted cuttings) are employed, at least one year old materials should be given preference.

On the other hand, for budded or grafted materials, those which have grown for two years should have to be used. This is because, budding or grafting is done when the rootstock has become one year old and after the operation has been done, it is also necessary to maintain it in the nursery for another year and thus, it is ready for planting in the orchard in the second year. Such a long period of two years may however, be shortened in some fruits plants by adopting suitable procedure, notably *epicotyl-grafting*. Use of plant growth regulators in proper form, concentration and stage of maturity also enhance the growth of the materials in all the above cases and thereby, render the materials ready at earlier than the normal age. Above all, micro-propagation system which of late, has become very popular for banana, apple and many other fruit trees employs suitable techniques in culturing the tissues for which, the materials could be made ready in shorter time.

The cost of a propagating material goes higher with advancing of age, which is because of the fact that regular watering, weeding, manuring, plant protection measures and its maintenance in the nursery involves some expenditure. Therefore, many orchardists prefer to purchase propagating materials at early age of them, i.e., when they have not become ready, which is with the idea that those could be purchased at low cost. Such a tendency could hardly be supported in the light of the fact that survival of the under-aged materials becomes difficult after planting. Even handling of such materials may cause damage to them.

On contrary, many orchardists prefer to procure higher aged propagating materials paying even high price with the idea that those would produce fruits at relatively short time. Although it is true that higher aged materials come to bearing early, survival of those after planting has great difficulty. If such materials are at all to be used, they should be planted with sufficient roots which means that very large earth-balls have to be used. It is also necessary that they should be in properly pruned condition.

When higher aged materials are used, a better system that may be taken up in the nursery is *repeated transplanting*. This implies that they should be transplanted from one bed to another 2 or 3 times at interval of 1 to 2 months. By this practice, although the seedlings will age but they will not become very tall. On the other hand, their stems become thicker, which is a desirable character. Another merit of repeated transplanting lies in the fact that the plants are *hardened*. That is, they become more sturdy and are capable to resist adverse conditions to a great extent. In attaining this attribute, they can withstand the *shock* due to transplanting in the orchard, such that the rate of survival and subsequent growth after transplanting become increased. The vegetative growth of the propagating materials may also be brought under control by judicious application of plant growth retarding chemicals as listed in Chapter 14 (Ref. 14.2-15).

7. Grafting in Proper Manner

If grafted materials[5] are used for propagation, these should be checked to confirm of the following facts.

(i) The rootstock which has been used for a material should be well-suited to the soil condition of the orchard where it will be planted.

(ii) The graft has been separated from the mother plant at least a month before it is procured.

(iii) The graft union should be sufficiently strong, having no weak attachment. As shown in Figure 16.2 (a) there should not be any formation of *callus*[6] growth at the stionic[7] junction, which makes the graft-union weak.

5. Grafting: refers to bud-grafting (syn. Budding), or scion-grafting (syn. Grafting)

6. Callus: Simple parenchymatous cells arranged side by side.

7. Stion: A portmanteaue word, i.e., contraction of the words, stock (syn. Rootstock or Understock), over which grafting is done and scion (syn. Scion or Cion), which is grafted over the rootstock..

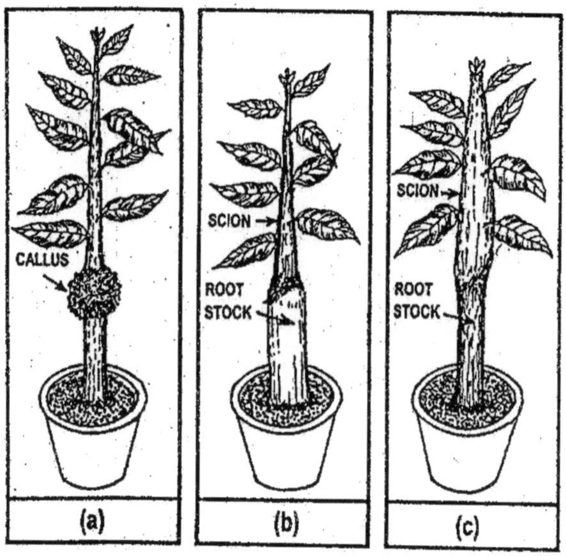

Figure 16.2

(a) Callus Growth at the Stionic junction; (b) Rootstock has Overgrown the Scion; (c) Scion has Overgrown the Rootstock.

(iv) It should be noted that grafting at very low level on the rootstock is also not desirable. This is because, such a material where grafting has been done near the ground level when planted in the orchard and cultivated, the irrigation water may come in contact with the graft-union and for this, the graft-union may be infected with disease organisms, weakened, or the scion may even be separated from the rootstock.

(v) The rootstock and the scion should have proven *compatibility*, which means that they should match together properly. Accordingly, as shown in Figure 16.2 (b) and 16.2 (c), neither the rootstock should *overgrow* the scion, nor the scion overgrows the rootstock.

8. Particulars of Propagating Methods

The propagating material should be procured from such a nursery that it should be able to provide detailed information of the material with correctness. At least the following informations are essential to be furnished.

(i) Variety of the propagating material.

(ii) Mother material from which propagation has been done and where it is growing.

(iii) For seedlings such as in papaya, whether the seeds have been procured from fruits developed by open- or controlled pollination.

(iv) For rooted cutlings (= cutlings) and layers (= layerlings), what type of materials have been used and from which mother trees and what method has been followed.

(v) Where grafted materials are taken, it is essential to know details of the rootstock and the scion materials that have been used. Besides, the method and the time of grafting should also be made known.

(vi) Whether any manure, fertilizer, plant nutrient salt, plant protection chemical, plant growth regulator or any other chemical has been used for the material and if so, the name of that, amount used, time of application and the purpose.

(vii) Age of the propagating material.

(viii) Whether similar kind of material can be supplied by the nursery in future, if mortality of what has been supplied takes place.

As a matter of fact, if the orchardist places order to the nursery beforehand hand and instructs to prepare the propagating materials as per his directions, the informations would be known to him. But if the materials are purchased without such a previous contract, the informations are essential to be known.

16.3 Pretreatment Before Planting

For encouraging vigorous growth of the propagating materials after they are planted in the orchard, some of the treatments to them done before panting are found to be much helpful. A number of *pre-treatments* are made to the propagating materials by the orchardists and some common ones are stated in the following.

1. Planting in Secondary Nursery Bed

It has been mentioned earlier under the Heading 16.2 (1) that the propagating materials are supplied by the nursery, maintaining the roots in earthen balls or in pots. The harmful effect of maintaining a material in a small earthen ball or in a small pot has been indicated.

However, even if the earthen ball or the pot is of large size, growth of the plant becomes slow after planting it in the orchard. As a matter of fact, the plants are maintained in the nursery in earthen balls or pots for a considerable period. Due to such confinement of the roots in pots or earthen balls, free growth of them both laterally and vertically is obstructed for not having sufficient space. Accordingly, the roots become *inward curled*. These roots remain curled for a long time after planting and as a result, the above-ground parts of the plants also do not make vigorous growth. To avoid this, it is necessary that before planting, the roots should be induced to direct and tend to *move outwards*. An effective method to do so is, planting the materials in a *secondary nursery* and to maintain in this nursery for some period before actually planting in the orchard. This practice greatly helps in straightening the ball-bound or pot-bound, curled roots and make them grow freely on all directions.

Beyond doubt, special care should be taken to prepare such a secondary nursery bed. The nursery bed should be located in partial shade and to prepare it, a trench should be dug (Figure 16.3). The depth of the trench should be 40-50 cm and the length and width may be as desired. It is better to provide a layer of sand at the bottom to 2-4 cm. Over the sand bed, a mixture of equal parts of loamy soil, well-decomposed leaf-mould and sand with little amount of wood ash and *neem cake* should be filled upto about 15 cm and over this layer, a mixture of loamy soil and well-decomposed leaf-mould in same proportion should be placed upto the ground level. Organic manure like compost or farmyard manure should rather be omitted as these may cause termite problem. The trench thus filled up, should be watered for few days. Some depression is likely to result due to watering and this should be filled upto the ground level with mixture of the top layer.

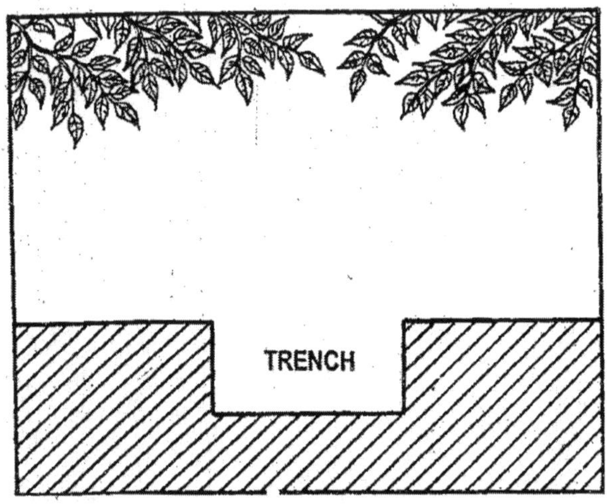

Figure 16.3: Trench made in Partial Shade.

If the material is maintained in a pot, it should be taken out from the pot taking great care that the soil remains *intact* and no part of the soil gets separated. This may be done by hands, if the pot is small-sized. To do so, as shown in Figure 16.4 (a), the pot should be held with one hand and the other handbe placed on the soil surface of the pot in such a manner that the forefinger and the middle finger are placed on both sides of the stem of the plant. Holding the pot in this way with two palms, it is turned upside down as shown in Figure 16.4 (b). When hung, the pot and the soil should be held very tightly. Now, while the soil of the pot is held with one palm, the other palm is taken off from the pot and as shown in Figure 16.4 (c), the side and bottom of the pot should be stuck with this palm lightly. On doing so, the entire soil will come out from the pot and will lie on the palm, which is shown in Figure 16.4 (d). (The operator should hold the pot and the plant at his own convenience.)

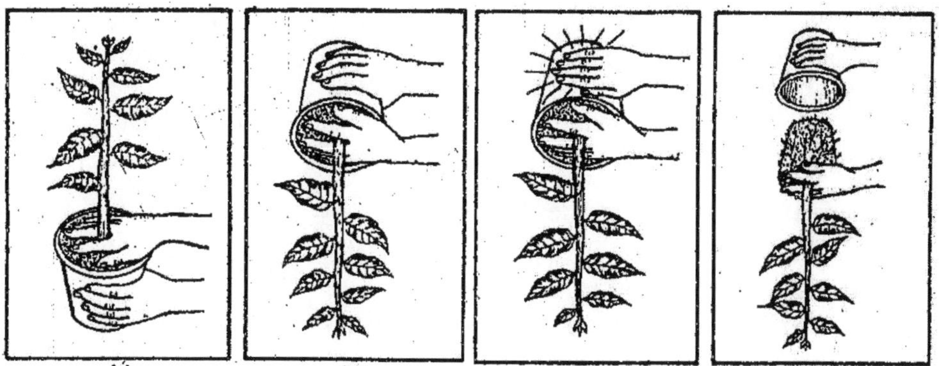

(a) (b) (c) (d)

Figure 16.4

(a) Pot is held with a palm and soil layer with other palm; (b) Potted plant is turned Upside Down; (c) Pot is Struck with Palm; (d) Soil and Plant have come Outside Pot.

Although the above method appears to be simple, it can be done by an experienced person only. It should also be kept in mind that the method is practicable only when the pot is of such size that it can be held with one palm. Moreover, removal of the entire soil without breaking it may be difficult if the soil of the pot is wet. Hence, watering of the pot should be stopped for a day or two and only sprinkled to facilitate the operation.

In case the pot is large and weighty there is great risk to adopt the method stated above. For such a pot, it is better to place the pot inside a polythene bag and is rested on ground. Then the pot is gradually broken by hammering lightly over the polythene bag from outside. When broken, the pieces are one by one taken out from the bag. But in this method also, there is chance of detachment of some soil and hence, some orchardists prefer another method to bring out the soil from the pot. They plant the propagating material in the secondary nursery bed along with the pot at first. That is, the soil is hollowed to make a scoop. Then the pot is carefully placed in that hollowed part. After that, the pot is lightly struck with a mallet to break into pieces. When broken, the pieces are removed from the scoop one by one and the soil is pressed taking from sides. By this method, even if any soil part in the pot is separated, it is brought to original position in embedded condition.

Watering of secondary nursery bed should be done lightly and frequently. Sprinkling is better than flooding and to do so, the plants should be bathed and the water sprinkled on soil. For the first few days, watering in this manner thrice a day is advisable. Subsequently, light irrigation may be done twice a day. The plants should be maintained in this nursery for a month or two and then transplanted in the orchard.

A decoction made from tobacco, turmeric, *neem* leaves, bayleaf and black cumin seeds may be sprayed occasionally to safeguard against minor insect pests. These materials should be boiled in water, cooled, strained through cloth, suitably diluted and finely sprayed. If aphids or other insect pests are still noticeable, necessary plant protection chemicals should be applied. Normally, diseases do not appear where there is no drainage problem but if observed, should be controlled by application of suitable fungicidal chemical.

2. Maintaining Optimum Leaf Area

Presence of leaves in a propagating material of a tropical or a subtropical fruit crop is always desirable. The leaves greatly help in establishment of the material after planting and also further growth of it. In fact, *leaf is the factory in the plant body* where the necessary components required for growth as well as formation of new organs in the plant, such as carbohydrate, protein, other compounds, plant hormones *etc.* are largely synthesized and then these elaborated synthates are translocated to other organs as and when they are needed.

On the other hand, in a leafy plant, leaf is the principal organ through which water is given off in the atmosphere in vapour form, *i.e.*, transpiration and evaporation and in some plants as water droplets, *i.e.*, guttation. The water vapour or water droplet that is given off by the leaves is filled up again by taking up water mostly from soil which is done by the roots and specifically speaking, the root-hairs, which are the absorbing organs for water as well as nutrients.

Now, if roots are less and leaves are more, the evopo-tranpirational loss of water through leaves will be greater than uptake of water by roots. If this is continued for some period, evidently the plant will suffer from water starvation. Accordingly, this deficit of water in the plant body will cause its drying and ultimately death of it.

Such an unwelcome situation can be averted, if there is a good balance between the roots and the leaves. Therefore, in case a propagating material contains many leaves and the quantity of roots is much less in it, *i.e.*, the roots are maintained in an earthen ball or a small pot, it is necessary that some of the leaves should be cut off. An important point to be remembered in this context is, trimming of some leaves so as to reduce the number should not be considered as the only condition. As a matter of fact, it is the *surface area* of the total number of leaves that is of concern and not the number of leaves only. Thus, if a plant has a large leaf area but number of leaves lesser than one which is just the opposite, it may be that water loss may be greater in the former. Therefore, which and how many leaves are to be trimmed is a matter that should be judged very carefully, keeping in view the number and size of leaves and the amount of roots. For plants having large leaf-blades, it is a better approach to trim part of a leaf only retaining the rest part intact, rather than complete removal of it.

Under the conditions of India, the mortality of the propagating materials after planting is noted to be due to the two principal reasons. These could be stated as follows:

(i) Irrigating heavily, or high rainfall with inadequate drainage.

(ii) Improper balance of the laminar area and the aborbing organs.

In fact, large number and large-sized leaves are usually encouraged in a propagating material by the nurserymen to attract the customers and sometimes it is even induced by them with application of some chemicals and plant growth regulators in solution form.

3. Trimming of Inflorescence

In case of vegetatively propagated plants, sometimes inforescences are found to be present, or these may come up within a few days or a month after it is procured from the nursery. This is commonly noticeable in a mango graft. Such type of protrusion of a flower in a material is due to development of the flower-bud that was already present in it while it was in the mother plant, *i.e.*, before it was detached from the latter. Further, vegetative growth that is required to produce flower-bud has also been accomplished in it before detachment from the mother plant.

Many orchardists as also common people consider it a desirable feature for they think that the plant is now going to give fruits. But this is not correct. These flowers produced by a plant in premature condition of a plant may give fruits but these fruits do not develop to proper size and do not attain desirable quality and hence, are of no value. On the other hand, presence of a flower in such a young plant hinders vegetative growth of it. (Vegetative growth and reproductive growth are antagonistic !)

Two important points are in fact, to be remembered in this connection. These are stated below.

(i) As much vegetative growth as possible should be encouraged in a plant till it attains the *maturity of flowering*. It should be remembered that higher the vegetative growth of a plant until it comes to the flowering age, *i.e.*, bearing maturity, higher will be the productive capacity of it subsequently.

(ii) *Vegetative growth and reproductive growth, i.e., flowering* have opposite trend. This means that if reproductive growth is there, vegetative growth will be hampered and *vice versa*. With consideration to this phenomenon, what desirable is, to trim off any inflorescence, flower or flower-bud whenever visible in such a young plant and thereby, allowing it to make vigorous vegetative growth in an unhampered manner.

4. Proper Training

Sometimes it is noticed that the branching of the propagating material in case of a tropical or subtropical fruit is not good. The following defects are usually noticeable.

(i) The plant is too tall with lesser branches or with no branch.

(ii) Profuse branches are there.

(iii) Branches are long.

(iv) At some parts, there is over-crowding of branches while in other parts, branching is sparse.

(v) Crotch angles (= Angle between two adjacent branches)are not good and so on.

In fact, such type of propagating materials should not be purchased to set up a new orchard. Nevertheless, if such undesirable features are noticed or happen to take place in a material planted in the secondary bed, that needs to be corrected. In order to correct these defects, in other words, to induce proper height and branching, what necessary is, *training*. Training is an operation done to a plant to impart any size, shape or form as desired. Evidently, it should be done at young stage of a tree, *i.e.*, before it has taken a definite shape and a strong framework. For erect growing fruit trees, the different methods of training are the same as have been described under the Chapter 9 (Windbreak rows). Among these methods, which one is to be adopted is a matter of judgment and it is to be decided in consideration to a number of factors. Above all, species of the plant is a very important factor and for some species, presence of branches or to induce those by the process of de-topping (Figure 9.10) at very early stage is not desirable.

Training is a skilful horticultural operation which is done not only in young stage of a fruit tree, but is also carried out as a continuous process during the first few consecutive years of growth and in phases. Further, a *mechanical shock* is produced in the plant when a branch of it is cut off and the plant may even die if the shock is heavy. In accordance with, it is advisable that for propagating materials of most of the fruit trees and especially those which are delicate, training by cutting a branch should not be attempted till those are established in the soil, *i.e.*, within a month after planting either in the secondary nursery bed or in the orchard. If any branch, regardless of thicker or thinner is cut off, application of a fungicidal paste over the cut surface just after cutting should also not be forgotten to disallow entry of pathogenic organisms through the cut surfaces.

5. Application of Nutrient Salts, Plant Growth Regulators *etc.*

Foliar spray with aqueous solutions of plant nutrient salts over the propagating materials is found to make them sturdy and vigorous. Accordingly, their survival rate and growth after planting in the orchard become high. Some plant growth regulating compounds are also found to be effective to make improvement of the plants. But effectiveness of these compounds depend much on concentration, *i.e.*, they show positive results when applied in low or even in extremely low concentrations. Foliar application of the solutions of plant nutrient salts or the plant growth regulators even in slightly higher concentration than optimum may render great harm to the plants and may even kill them sooner or later.

In general, foliar spraying the propagating materials with aqueous solutions of 0.05-0.1 percent zinc sulphate, 5-10 parts per million of Indole Butyric Acid and 0.5-1 per cent common sugar (sucrose) are found to be effective. Each of these should be applied separately leaving a gap of 4-6 days. Spraying may be done twice or thrice at interval of 10-12 days.

16.4 Curing

In some plants and such as in pineapple plants, a treatment known as curing appears to be necessary for the suckers or the slips, which are largely used as the propagating materials. This consists of two steps, which are as follows.

(i) Cutting off the lower leaves upto 2-4 cm on the stem and cleanly trimming the lower end with a suitable knife at first.

(ii) After that allowing the materials to get partly dried on exposure to air.

There is in fact, difference of opinion in regard to both the length of cutting and the period of exposure. Many consider that stripping basal leaves for more than 2 cm may hamper establishment of the material while others opine that it should be upto a length of 6 cm or more for better result. Again, exposure to air has been advised by many as not more than 5 days, as opposed to others who advocate at least 10 days. In Hawaii, it is also a practice done by many pineapple growers to expose to air for several months in single layer.

What should be the length of the basal stripping and air-exposure depend in fact, on a number of factors, notably, the varieties or types, maturity, health, length of the materials and the leaves present, climate and soil condition, whether the crop should be grown in full sunlight, shade or semi-shade and the method of culture to be adopted. In any case, the operation is known to be of great help at least for the following two reasons.

(i) Higher production of rots at the early stage, *i.e.*, after planting, for which the material is easily established in the soil and hence, is of great help especially in dry condition.

(ii) The materials are not attacked by leaf rot disease which may result in case of uncured suckers, when planted in moist soil.

Chapter 17

Planting of Propagating Materials

17.1 Principles

Whatever genuine and perfect the propagating materials are, the orchard cannot be set up in a scientific manner, if there has been any error in the operation of their planting. *Planting* does not mean just infixing the propagating materials arbitrarily on the orchard land. It should be noted that plentiful considerations are in fact, needed to be kept in view as regard planting is concerned in the orchard.

Among other conditions, serious attention is needed to be paid on two aspects in regard to planting. These are, the *seasons* of planting and the *methods* involved in the planting process. These have been elaborated under the following two Headings.

17.2 Season of Planting

Prof. W.B. Hayes of the erstwhile Allahabad Agricultural Institute quoted, " The season of planting is a large factor in the successful establishment of the orchard." (Ref. Hayes, W.B. 1970. *Fruit Growing in India*. Kitabistan, Allahabad).

In the practical sense, a number of considerations in fact, become necessary to be made to determine what should be the actual season when propagating materials are to be planted.

To get satisfactory answer to this query, a number of considerations need careful attention. These could be stated as follows:

(i) The species of the fruit crop.

(ii) Variety(ies) or type(s) or hybrid(s) of the species.

(iii) Climatic condition of the locality.

(iv) Soil condition of the orchard site.

(v) Irrigation facility.

(vi) Post planting care to be taken.

(vii) Whether earliness of fruiting is desired.

(viii) Funding capacity of the orchardist.

(ix) Other factors.

1. Planting of Tropical and Subtropical Fruit Plants

In tropical climate, the propagating materials that come under tropical or subtropical fruit crops may be planted at any season of the year. Precisely speaking, where tropical climate is more clearly pronounced, such that rainfall is high or very high, distribution of the rainfall is widespread and the winter is mild, *i.e.*, indistinct, planting may be done at any time of the year. Such type of climate is largely met with in the *coastal tropical area.* However, if the tropical atmosphere is not so distinct in such a torrid zone, although planting may be done at any season but it is advisable not to do so at least in the winter season. In this connection, the following aspects have to be kept in view.

High *meristematic activity* is necessary for protrusion of new roots and new shoots and their growth. Meristem is a localized group of cells. They are in actively dividing (cell-division) stage in plants from which permanent tissue is derived. The meristematic activity of a plant is maximum in the rainy-season and spring. This means that protrusion and subsequent growth of the shoots and roots are of highest extent in these two seasons. Now, when the propagating materials are planted at the time of having maximum growth activity of them, *i.e.*, in rainy-season or spring, they produce leafy shoots at the above-ground and rootlets at the under-ground parts quite freely.

The higher number of roots and rootlets produced by the propagating materials in fact, bind the soil more firmly than when the roots are lesser in number. As a result, the materials are *firmly anchored in the soil*. Besides this, higher production of rootlets obviously give rise higher production of *root-hairs*. The root-hairs are the actual organs in a plant which take up water and nutrients in it from the soil. Accordingly, metabolic activity in the plant body is also increased and this results in quick and firm establishment of it in the soil medium where planted.

In addition to the above fact, production of new leaves is also an important factor for establishment of the plant. This is because, the new leaves are known to produce some hormonal and other chemicals. These compounds which are synthesized in the leaves are translocated to the roots and there, they provide the *stimulus* for higher production of roots. Higher production of roots has in fact, a greater grasping effect of soil and thereby, the plants get established in the medium in a relatively short time.

Between the rainy-season and spring, the season which is more favourable for planting depends in fact, on the following factors.

Planting in rainy-season has the greatest advantage in that, artificial irrigation is not required regularly after planting, except when there is a gap of rainless period. High relative humidity prevailing in this season also greatly reduces the loss of water from the plants due to evaporation and transpiration and as a result, requirement of water is lesser as well as growth of the plant is increased, Besides these, the materials planted become well-established before the onset of winter and hence, these are less prone to cold injury, if there is a sudden drop of temperature in the winter season. Above all, the plants are also not damaged due to hot-weather in summer as they become firmly established by then.

Despite the above advantages, rainy-season planting has a great drawback as regard water-logging. Heavy rainfall if occurs, may cause stagnation of water around the plants and thereby, results in weakening them and also disease infection, especially the collar-rot. Water-logged condition takes place where drainage of the soil is poor as occurs in clayee soil, while rainfall is high. Poor drainage in the subsoil layer due to impervious material is rather more harmful and causes suffocation of the roots. Therefore, the advantage of rainy-season planting is obtained where drainage facility is ensured as is the case in loose textured, porous soil. However, if planting in rainy-season is at all to be done, the orchardist should follow the general guideline, *i.e.*, in low rainfall area, planting should be done at the beginning of rainy-season and in heavy rainfall area, it should better be delayed till the later part or end or rainy-season to avoid any possible risk.

Planting in the spring season obviously requires artificial irrigation regularly after planting as there is no rain. Irrigation after planting should in fact, be done more frequently and with lesser amount and should be continued in this manner for all the summer months, till the onest of rainy-season. Therefore, cost of irrigation goes high for planting the propagating materials during this season. Nevertheless, spring planting is preferred under certain conditions in many parts of India and for a number of reasons.

As a matter of fact, the actual season of planting of propagating materials for tropical and subtropical fruit crops is judged by a number of considerations, which are, species and variety of the propagating material, rainfall and climatic condition of the locality, drainage and other soil conditions, method of propagation followed, other facilities available and funding capacity of the orchardist. For example, mango grafts in all over India are known to be planted in the rainy-season and same is the case perhaps with some citruses, guava *etc.* Banana is usually planted in spring under subtropical condition which is with the idea that the plants get the advantage of better growth in this season and in summer. In Kerala also, planting is usually done in this season while in the west-coast of India, planting of banana is done most commonly in September-October and also in the months of April or July-August. Under the conditions of West Bengal and eastern India, banana is planted from spring or in rainy-season if the suckers are small and in autumn when the suckers are somewhat old.

Similarly, planting of papaya which has a great impact on time of flowering is done in the eastern India usually at the end of rainy-season or extending upto the month of November while in most parts of south India, it is done in spring. Grape is planted from January to March under subtropical situation and litchi is planted in the Punjab from August to November unlike in full rainy-season as is commonly done in major parts of northern Bihar and northern West Bengal. For planting date in the Punjab, September is considered better than the spring. The months of September-October are also considered superior for planting loquat in the northern India. In case of fig, early spring is considered suitable for planting in the western India. For the tropical fruit tree, jackfruit planting from June to September is preferred where there is south-west monsoon and during the months of July to December in most other areas of India where this fruit tree is grown.

2. Planting of Temperate Fruit Plants

The propagating materials of most of the temperate zone fruit plants are planted in the winter season. The reason is, they are in *dormant* condition in this season, which means that they do not have any growth and do not contain any leaf as those are shed. They are handled with bare roots at this season, *i.e.*, not covered with soil. As a result, the plants are not damaged due to low temperature in winter. On the other hand, the vegetative buds that remain present in the plants are benefited on receiving the low temperature in winter. In fact, the buds of temperate fruit crops do not sprout and give rise to new growths unless they receive low temperature in winter for at least a specific period of time.

On receiving this *chilling exposure* in winter for the required period during their dormancy, the buds sprout and start growth when the winter is over, *i.e.*, with the advent of spring. With sprouting of buds and growth of shoots, growth of roots also takes place and thus, within the spring season or shortly thereafter, the plants become established in the soil where they have been planted. It should be noted in this connection that varieties under some species of fruit crops are there, whose chilling requirement is not so acute and thereby, they are able to grow satisfactorily in warmer temperate zone or in the subtropics. For these varieties, exposure to low temperature for bud-break though becomes necessary, but very low temperature is not required. But those varieties and species which are acutely adaptable to temperate climate and grow in very high altitude in India have absolute requirement of low chilling for bud-break.

However, although planting in winter theoretically gains support, it has practical limitation where winter temperature goes extremely low. In such places, the newly set young plants actually suffer from *low temperature injury* and even may be killed. Therefore, such areas having very low winter temperature, planting is in fact, delayed and should not be done till the end of the winter season or even in early spring to save the plants against severity of winter. This adjustment of planting season in accordance with local conditions becomes necessary for the temperate fruits, nuts and the berries.

Apart from winter, planting in autumn or in rainy-season is also done for some temperate fruit plants like pear, raspberries, hazel- nuts *etc.* under certain conditions. For strawberry, planting is done in the months of March-April or September-October in the hills, while in January-February when grown in the plains.

17.3 Methods of Planting

The method of planting in an orchard or in a plantation of fruit crops depends on a number of factors, important of which are plant species, condition of the soil such as texture, depth and fertility,planting systems, rootstocks used for the materials, method of culture to be adopted, choice of the orchardist and so on.

Before going to describe the methods, it is pointed out that except in case of the method stated below under the Heading, 17.2 (6), *i.e.*, where seeds are directly sown, in all other methods, the propagating materials are planted in the allocated positions of the orchard. Now, if the materials are maintained in the secondary nursery bed as stated in the Chapter 16 under the Heading 16.3 (1), the propagating materials are needed to be uprooted from the nursery bed for planting. This operation, *i.e.*, uprooting should be done with utmost care that the roots are least injured. To ensure this, the moisture in the bed should be at desired level, *i.e.*, neither high nor low and such a soil-moisture condition should be maintained by regulating the irrigation in the bed before two to three days of uprooting. While doing uprooting, it is no doubt, essential that sufficient soil is lifted along with the roots of the plants so that they remain well-covered in the soil and are not exposed.

The different systems that are commonly followed in planting propagating materials in the orchard or elsewhere have been presented in the following.

1. Filled-in Pits

(a) Principle

The fruit plants in their young stage are very delicate. Hence, it may be difficult for them to get established and take good growth during primary stage, if they are planted as such in the ordinary soil of the orchard. But vigorous growth during young stage has beyond doubt, an essentiality. Accordingly, to boost up growth of the fruit plants during their early stage, it is necessary that they should be planted in such a type of soil which has good texture, high fertility and free from injurious pests and disease organisms. Making the entire orchard land into such a condition is difficult and expensive. Besides, it is not necessary also due to the fact that during the primary stage of growth, the fruit plants do not send forth roots to long distances.

On the other hand, if the small area of the soil upto which the plants are expected to spread their roots during first few years of growth are brought to such condition, the purpose may be served. With this objective, it is recommended to *dig pits* in the orchard (Figure 17.2) at position where plantings are actually to be done and then fill the pits following the method as described below. This is done so that a *congenial medium* is prepared to encourage better growth of the propagating materials.

(b) Depth of Pits

The pits are to be dug at positions where the propagating materials are to be planted. What should be the depth and diameter of a pit is however, a controversial subject among the fruit scientists. Thus, a section recommend large and deep pits, for according to them, larger and deeper the pits are, benefits rendered to the materials that will be planted there on filling is also greater. They advocate such deep and wide pits with the idea that as improvement has been made in the texture and fertility of the soil over a deep and large area, the roots of the propagating materials planted there on filling the pits will grow vigorously in an undisturbed manner not only in the first year but also even few years after planting.

On contrary, such a viewpoint is not supported by the other group. They are of the opinion that such type of large and deep pits are in fact, unnecessary and beyond doubt, expensive as well. Prof. W.B Hayes in his book has quoted, " It is rather advantageous to bring the whole field into good condition before planting, instead of digging large pits and filling them to enrich the medium.

As a compromise, the following considerations may be kept in view.

(i) By and large, a pit which is 50-60 cm deep and 60-70 cm wide appears to be fairly satisfactory for planting propagating materials in the orchard for most of the common fruit trees and shrubs.

(ii) For shallow-rooted fruit plants like papaya, phalsa, fig, custard apple, litchi *etc.*, even smaller pits may be made.

(iii) Pits having greater diameter than depth are of more benefit to the fruit plants.

(iv Where soil is inferior, the entire land should be prepared at first by manuring, ploughing and levelling and then, 50-60 cm cubic pits may be made.

(c) Use of Planting Board

The pits are to be dug up in places where propagating materials are actually to be planted, *i.e.*, where the pegs that indicate planting positions (Figures 12.5-12.9) had been infixed (Chapter 12). Now, while digging the pits, the planting position indicating pegs obviously, need to be removed. Due to that, the exact locations where the pegs were fixed could be difficult to be ascertained at the time of planting by eye-estimation only.

The above difficulty can be overcome and the exact planting positions be known by the use of the tree-setting *planting board.* As shown in Figure 17.1 (a), the planting board is only an ordinary wooden plank of convenient size and may be about 1.6 metres long, 20 cm wide and of about 2 cm thickness. It has three V-shaped notches, one of which is in the centre and one at each end. (The planting board may be of various designs in providing the notches, apart from the one described here.) To work with the board, it is brought near a fixed peg and placed on ground in such a position that the peg fits into the central notch. Without disturbing the board from this position, two pegs are firmly fixed to the ground at the other two notches of it. The two pegs thus, fixed along the notch at each end of the board are called *guide-pegs* [Figure 17.1 (b)]. On fixing the guide-pegs, the planting board is removed from this place. But the guide-pegs which are fixed will remain there till planting of the propagating material is done. The central peg which was already fixed is called the *tree-marker* peg and this is now pulled out to dig the pit at this position.

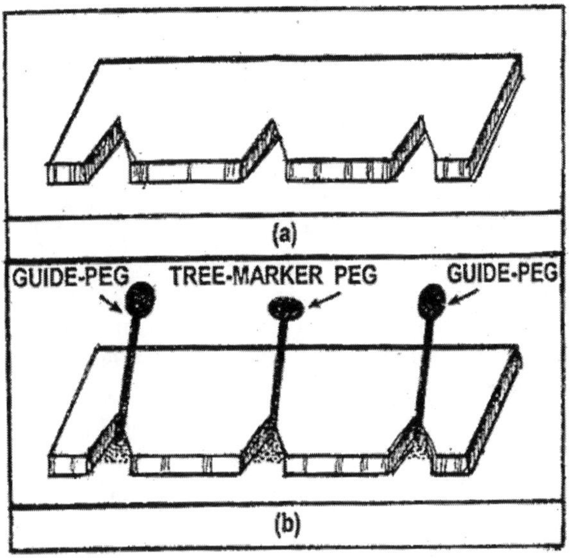

Figure 17.1: (a) Planting Board; (b) Fixing Guide-pegs.

(d) Making Pits

The pits are made by digging up the soil with spade or any suitable implement. The soil dug up from a pit should be placed on the ground close to the pit. This should be done in such a manner that the soil dug up from the upper layer of the pit is left at one side of the pit, while the soil dug up from the lower layer of the pit at the other side of the pit (Figure 17.2).

(e) Weathering of Dug up Soil

The soil dug up from each pit and left on the ground needs to be *weathered*. This means that the soil should be left as such, *i.e.*, exposed to air for a certain period, *i.e.*, days. Weathering of soil is in fact, done with the idea that the eggs and larvae of insect pests, disease organisms *etc.*, if present in it are destroyed and that the soil is aerated. In this way, when the soil has been adequately weathered, it is returned to the pit to fill it up.

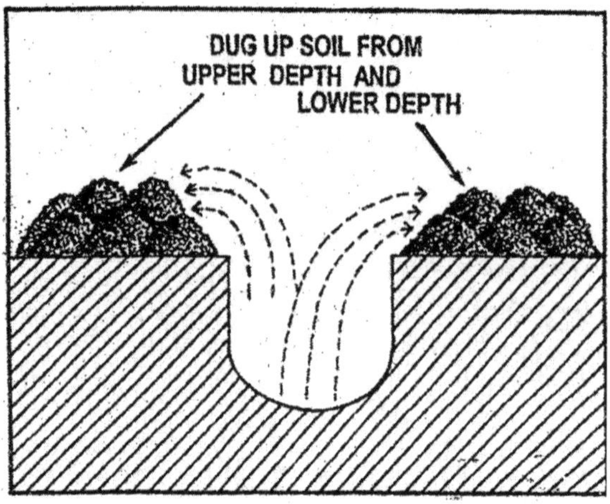

Figure 17.2: Planting Pit and Heaping Dug up Soil on Ground around it.

How long the soil should be weathered is however, a topic that has two opposite viewpoints. One group recommends long weathering upto a month or more as it ensures complete destruction of insect eggs, larvae, nematodes, disease organisms *etc.* The other group is not in favour of such long weathering process, because according to them, much of organic matter and nitrogen content of the soil are lost by this long exposure. The opinion of the latter group is however, not supported by the former group on the ground that the dug up soil is in fact, made artificially enriched by mixing with manure and fertilizers before returning to the pit. Hence, even if the soil loses organic matter and nitrogen, that are actually recompensed. The opinion of the latter group perhaps gains support and is acceptable by many orchardists.

Accordingly, it may be suggested that if planting is desired to be done in the *rainy-season*, the pits should b dug in *summer*. By this, the dug up soil is kept exposed to scorching heat of summer and thereby, becomes virtually free from pests and diseases. After such long weathering for a month or so, the pit is filled with the weathered soil at the beginning of the rainy-season. Similarly, when planting is to be done in spring in or winter season, pits may be dug after the rainy-season is over.

(f) Filling Pits

The propagating materials are delicate at their infant stage. Hence, it becomes difficult for them to take up good growth, if planted in soil of adverse nature. To get rid of this difficulty, improvement in the textural condition of the dug up soil and enrichment of its fertility become necessary before it is returned to the pit, in order that the roots of the materials are surrounded by soil of soft texture and increased fertility. This is accomplished by thorough pulverization of the soil and addition of suitable amount of well-rotten manure and fertilizers to it.

Apart from incorporation of manure and fertilizers, fortification of the soil with plant protection materials as safeguard measures against pests and diseases also seeks justification. Thus, the dug up soil should at first be prepared by pulverizing and mixing with manure, fertilizers and plant protection materials and then it should be returned to the pit.

What and how much quantity of manure and fertilizers are to be added to the dug up soil depend on a number of factors, notably the soil condition, its fertility, depth and diameter of the pit, duration of weathering, species and variety of fruit plant to be planted, season of planting *etc.* However, where specific recommendations are not available, it seems satisfactory to mix the dug up soil with 30-40 kg of well-rotten farmyard manure or compost, 1.5-2 kg of super-phosphate and 0.2-0.3 kg of muriate of potash or 2.5-3 kg of wood ash for a pit of about 50-60 cubic centimeters. The manure and fertilizers are to be mixed with the entire soil dug up from a pit, *i.e.*, the soil dug up from the upper as well as lower parts of the pits.

Inoculating the dug up soil with the bacterial fertilizer containing nitrogen fixing bacterium, *Azotobacter chroococcum,* if possible to be done is advisable to raise the level of nitrogen. Inoculation with phosphate solubilizing bacterium, *Bacillus megaterium* var. *phosphaticum* also offers advantage to enhance growth of the planted materials.

As regard addition of plant protection materials, thorough mixing with 5 per cent Aldrin or Heptachlor or 1-2 kg of *neem cake* to the dug up soil of each pit becomes necessary where there is problem of white ants. Many horticultural scientists even advise to sterilize the dug up soil following treatments as started in Chapter 7 under the Heading, 7.4 (1,2).

When the soil is prepared in this way it is ready for returning to the pits. But before the pits are filled, an important step should not be overlooked. That is, especially when the planting pits are dug before a month or more, a hard crust is very often found to form on the side walls of the pits. As a result, the roots of the planted materials may have difficulty to penetrate such hard-coated walls and hence, cannot extend freely outside the soil and manure filled pits. Therefore, as a precautionary measure, it is advisable to scrape out the walls of the pits by means of a hand-hoe (*khurpi*) or other implement.

After the walls of the pits are scraped, they are filled with the soil which has been mixed with manure, fertilizers and plant protection chemicals. Filling of a pit should be done in such a manner that soil dug up from the upper layer of the pit and is mixed with manure, fertilizers and plant protection chemical be placed at the lower part of it and over this, the soil of the lower layer which has also been incorporated with manure, fertilizers and plant protection chemicals be placed. After filling the pits in this way, these should be regularly watered. On doing so, the soil settles thoroughly in the pits. If the soil in any pit is found to be depressed due to watering, it should be made upto the ground level with the soil manure mixture before planting of propagating material in it.

(g) Planting

If the propagating materials are maintained in a secondary nursery bed (Chapter 16), those should be uprooted with utmost care so that the roots are least injured. It is also necessary that sufficient soil is lifted along with the roots of the plants while uprooting, such that they remain well-covered in the soil and are not exposed. For this purpose, the moisture content of soil of the secondary bed should be at the desired level. To ensure this, soil moisture should be checked regularly for few days before lifting the materials and watering of the bed should be done judiciously.

As in the case of nursery bed, the moisture content of the filled-in pits should also be at the optimum level, *i.e.*, neither high nor low. Accordingly, these should also be checked regularly for few days before lifting the materials and watering of them should be done with judgement. It may even become necessary to cover the filled-in pits with plastic sheets or leaves, if there is danger of water-logging during the rainy-season.

During planting, the planting board should be brought to the site of the pits. Besides this, trowels, hand-hoes *etc.* rope and strongly built sticks (may be of bamboo) also become necessary. Each stick which may be about one metre long should better be painted previously with coal-tar but at the time of their use, the coal-tar should be in thoroughly dried condition.

For planting, the soil of the filled-in pit should be scooped out with the trowel or the spade to make a hollowed part there in the centre [Figure 17.3 (a)]. Then 2-3 coal-tar painted sticks are firmly knocked in at close distance along the wall of the hollowed part, which is with the idea to provide support to the plant which to be planted there [Figure 17.3 (b)].

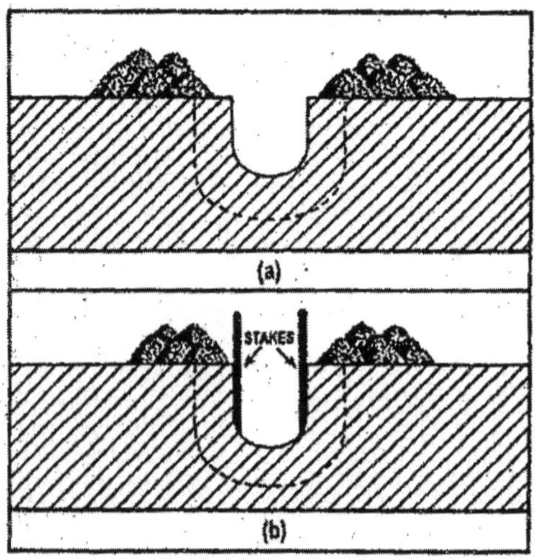

Figure 17.3: (a) Central part of the Filled-in pit has been Scooped out; (b) Stakes have been Infixed for Support to the Material when Planted.

The propagating material is which to be planted is then uprooted from the nursery bed with great care as stated above. In order to determine the exact position where planting is to be done, the planting board is now made use of. To use, it should be placed on the ground in such a position that the two guide-pegs which already remained infixed there [Figure 17.1 (b)] fit into the two notches provided at both ends of it [Figure 17.4 (a)]. On doing so, the position of the central notch is known and this is the position where the tree-marker peg [Figure 12.1 (b)] was actually fixed. Evidently, planting should be done there which means that the earth-ball of the propagating material should be placed in the hollowed part of the filled-in pit and at position where the central notch of the planting board is located [Figure 17.4 (b)].

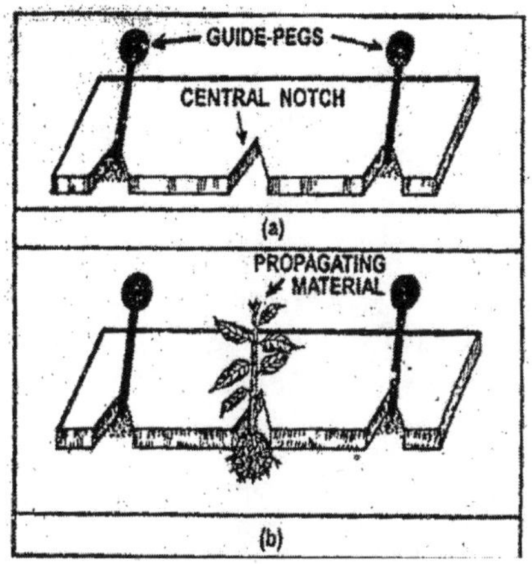

Figure 17.4
(a) Guide-pegs fit into two Notches at both ends of the Planting Board;
(b) Planting along the Central Notch of the Board.

Seemingly, the act of planting, *i.e.*, setting the earth-ball of a propagating material into the hollowed part of the pit as stated above appears to be an easy process. But in fact, complete success of the operation is difficult to be achieved unless heedful consideration is given on the following aspects.

(i) Adequate precaution should be taken that roots of the planting materials and especially, the tap-root if any, should not be twisted or broken, during planting. The growth of the plant may be stunted, if happens so.

(ii) If the propagating materials are planted directly on procurement from the nursery, *i.e.*, they are not maintained in the secondary nursery bed, their earth-balls should be somewhat loosened by giving pressure with hand or striking lightly with a mallet. This should be done with the idea that ramification and free growth of the roots outside the hard earth-balls are not impeded.

(iii) Care is also necessary that the material is planted vertically straight, *i.e.*, absolutely perpendicular and not in a tilted manner. For in such case, the tree may be lop-sided, or the lower part of it may not be straight when it will be in grown up condition. This will cause stunted growth, improper branching and poor performance.

Although an expert operator plants a propagating material vertically straight simply on eye-estimation only, minute observations sometime reveal that planting in an absolutely perpendicular manner has not been done for all the materials that have been planted. If there is likelihood in failing to achieve such type of accuracy in planting, it is rather advisable to use a suitable *plumb-line* to maintain the perpendicularity of the material during planting as narrated in the following.

To make the plumb-line, a small and flat piece of iron having a hole in it may be taken. A soft thread should be tied to this iron-piece through the hole. Now, as shown in Figure 17.5 (a), the other end of the thread should be tied to the upper part of the plant so that the plumb hangs freely. If there is danger of damage or breakage of the plant by tying the thread with it, one person will hold the end of the thread touching it with the upper part of the plant. In this way, when the plumb-line comes to rest after swaying side to side, it will indicate perpendicularity. In accordance with, planting should be done along the plumb-line and thereby, planting in a perpendicular manner could be ensured.

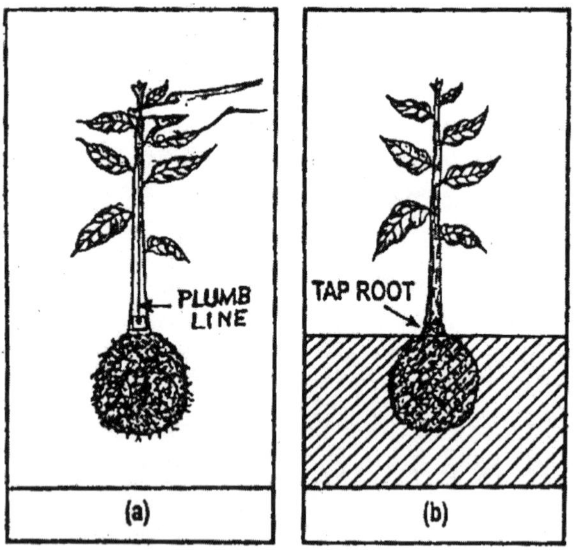

Figure 17.5
(a) Planting Vertically Straight using Plumb-line;
(b) Samll part of Tap-root is left above Ground during Planting.

(iv) An important point on which close attention should also be given is, *neither any part of shoot be buried underground, nor any part of the root be exposed above the ground* when the plant is established, meaning thereby that the junction of the root and the shoot should lie on the ground level. If this does not happen so, the *collar region* (= junction of root and shoot) of the plant may be disease infected, main root may be damaged and growth of the plant may be seriously hampered. However, even if the plant is set in this way taking utmost care, it may be seen after a month or so that some part of the stem has gone underground. This actually happens due to the fact that the soil with which the pit has been filled is made pulverized and mixed with manure. Now, when planting is done in such type of loose soil and thereafter, regular irrigations are given, the soil sinks down. As a result, part of the stem becomes buried under the ground.

As shown in Figure 17.5 (b), the above difficulty may be overcome if a *little part of the root is kept above the ground* during planting. In fact, this is suggested with the idea that when the plant will subside into the soil, the part of the root lying above the ground will only go under the ground and not the stem. Thereby, the root-shoot junction will come to the ground level. Beyond doubt, this should be done carefully that longer part of the main root is never kept exposed above the ground.

(v) After planting, care is also necessary that the plant is not tilted due to watering or with wind. For this purpose, it should be loosely tied to the coal-tar painted stakes which have been infixed close to it.

Nevertheless, the stakes may not always be enough to support the plants to stand vertically straight. In fact, regular watering and blast of wind may sometimes cause tilting of the plants, which may be just after planting or a few months later. To avert this, it is advisable to embed pebbles or brick-bats on the ground around the collar of the plant and also to heap those around the stem as shown in Figure 17.6. These materials may provide adequate support to prevent tilting of the plant. Heaping pebbles or brick-bats is also helpful to the plants in maintaining cool atmosphere to their root zone. But heaping these materials for long time is not desirable as it may be a breeding place for insect pests and even an abode of snakes. Hence, the pebbles or brick-bats should be removed out after a fortnight or a month.

2. Trench

In this method, terenches are dug at first like the *cattle-proof trench* as stated under the Heading, 6.4 (6) in the Chapter 6 (Figure 6.15). The dug up soil from the trench is weathered and then it is pulverized, mixed with well-rotten manure as also fertilizers and pesticides and then returned to the trench. The trench, filled-in with soil enriched with manure and fertilizers is regularly watered and if the soil is depressed, it is brought to the ground level with soil manure mixture.

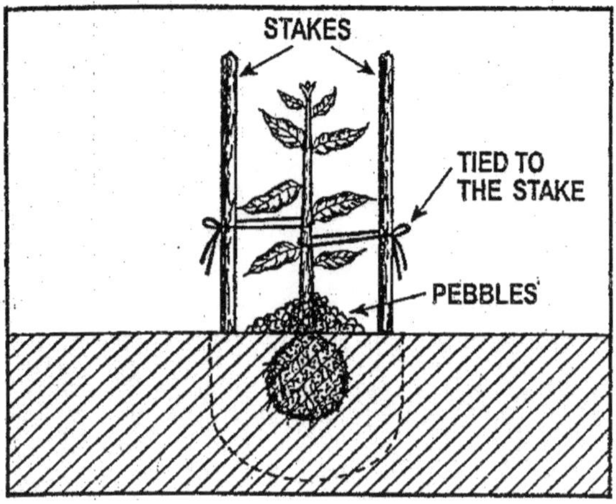

Figure 17.6: Plant has been Loosely Tied to the Two Stakes, besides Embedding and Heaping Peebles around its Base. (Instead of Two Fixing Three Pegs in a Triangular Manner give better result).

After that, planting is done in the filled-in trench. Thus, in principle, the method is the same as in filled-in pits. The only difference of this method with that of the pits is, instead of digging individual pits, long trenches are dug as is done in planting of sugarcane sets.

The trench is dug with spade and pick-axe, maintaining a depth and width of about 20 cm or little more and weathering of the dug up soil is usually done for a week or a fortnight only.

As a matter of fact, the method is adopted where plant-to-plant distance in a row is very short as in such case, instead of digging individual pits at close distance, digging along trench is convenient. The method has wide use to make a row of wind-break or hedge. For fruit crops, this is followed where hedge-row system of planting is done as in case of phalsa, pomegranate, ber, karonda, mulberry *etc.* and also in pineapple. In central part of West Bengal, some banana growers prefer this system for some varieties of banana. However, the method has little applicability for most of the common fruit crops.

3. Ridge

Making ridges and furrows by ploughing and with spade and planting on the ridges is also a practice done for some fruit crops in some parts of India. Planting on ridge is in fact, a very common system for pineapple under the conditions of Kerala. Some banana growers in the central part of West Bengal also adopt this system, especially when mixed-cropping of fruit is done, *e.g.*, banana, papaya and pineapple in alternate rows. Planting is done like hedge-row system and less space is maintained between rows. For other fruit crops, ridges and furrows are not done except for some short duration crops notably, cape-gooseberry which is grown like an annual crop. In some parts of northern India, Punjab and Haryana, jamun trees are sometimes planted on ridges. The ridges are alternated with thick growing but comparatively less tall trees such as ber or mulberry.

4. Raised Bed

Preparation of a bed raised to 15-25 cm above the ground level by pulverizing the soil and mixing thoroughly with required amount of manure and fertilizers is a common system for the culture of pineapple in the eastern States and also in many parts of south India. The width of a bed varies according to number of rows that are set and the length is maintained as desired. Paths of 30-50 cm are also provided between beds. For other fruit crops, the method is not found to be followed. However, a section of papaya growers in some parts of south India and east India prefer this method, perhaps as a safeguard measure against the collar-rot disease that results due to poor drainage.

5. Soil Mound

Making high mound of soil and planting fruit trees over that is a method which is done in some parts of India, under certain conditions. The soil mound is made, bringing the height to 1.5-2 metres or even more from the ground level and the top portion is made flat maintaining a width of 1-1.5 metres. The length of a mound may be 10-20 metres and a gap of 2-3 metres is left between two successive mounds. Spacing between two mounds depends on the fruit crops grown but not less than 10 metres. Thus, the mounds look like bunds over which fruit crops are grown (Figure 17.7).

Figure 17.7: Soil Mounds over which Fruit Trees and other Crops are Planted.

The method is particularly adopted in areas which remain inundated for a long period in every year and the level of water may raise upto a metre or more. Thus, the bunds are made to protect the fruit trees from water stagnation.

In some parts on the banks of the rivers, Ganga and Yamuna in north India which are locally known as *taahl* area and also on the river banks in other parts of India where the problem of inundation exists, some fruit growers resort to this method, although it is costly.

Instead of tall and vigorously growing fruit trees, dwarf and compact trees or shrubs are usually grown on the mounds. In the *Sunderbans* forest area in the coastal parts of West Bengal where the land remains flooded with tide-water for some hours of the day, some local agriculturists make such soil mounds to plant coconut, arecanut, bael, dwarf varieties of sapota, wild date *etc.* along with sugarbeet, cotton and vegetables. (Interestingly, the Royal Bengal tigers and pythons are also sometimes found to take shelter on the mounds during tide-water flooding but they do not attack anybody at that time.). Soil mounds may be made by mixing with stone chips and soil.

6. *In situ*

In this method, pits are at first made at desired locations in the orchard where the pegs were fixed (Figures 12.5-12.9). The dug up soil is weathered and returned to the pits with or without mixing manure and fertilizers. Then seeds are sown or rooted cuttings or layers are planted in the filled-in pits.

Sowing of seeds is however, more common in this method, the objective of which is to avoid transplanting. As a matter of fact, seedlings of many species and especially those having long tap-roots do not stand well and take up good growth after transplanting. Hence, to dispense with transplanting, seeds of these species are sown directly in the orchard. Instead of one seed, a number of seeds which may be upto 10 are in fact, sown at one place maintaining close spacing. Regular watering should be done after sowing and when do not make good growth they are uprooted. The remaining seedlings are allowed to grow for a week or a fortnight on applying irrigation regularly if necessary and then, those whose growth is not vigorous are again uprooted. The process is continued in this way, such that ultimately only one seedling which has made best growth remains present to grow and all the rest are uprooted.

As stated above, the method is adopted for such fruit crops which are needed to be propagated by seeds and which do not withstand transplanting well, e.g., papaya, cashew, jackfruit jamun, sometimes mango etc. However, raising seedlings *in situ* in the orchard is done for another purpose, *i.e.,* to prepare them as rootstocks for grafting scion materials over them following suitable method (budding or grafting by separation process). When the objective is this, uprooting of the seedlings at early stage should not be done. That is, all the seedlings are grafted with scion materials over them. After grafting, the plants should be allowed to grow with regular watering, weeding etc. When graft-union has taken place, i.e., after about two months, those grafted materials which do not show good performance should be uprooted and the rest are allowed to grow undisturbed. Again, after a gap of two months or so, uprooting should be done of those whose performance is relatively poor. The process should be continued in this way for six months to one year, so that ultimately only one grafted plant which has got through the screening test, in other words, which showed best performance all along is only left to grow and the rest are all uprooted. The merit of this system lies in the fact that the chance of loosening of the graft-union as sometimes happen when planting in the orchard is done after grafting is avoided. However, success of the method depends upon on a number of factors.

Apart from sowing seeds, planting of rooted cuttings or layers in the filled-in-pits is also done to make them as rootstocks. That is, when these plants are well-established in the soil, grafting of scion materials is done over them. When graft-union has taken place, uprooting is started in instalment as stated above.

Chapter 18

Post-planting Care

18.1 Principles

There is a strong desire of some people to take great care of a plant upto the time it is planted but they lose interest after planting is done. Thus, if the plant makes poor growth or even it dies soon after planting, they blame the health of the plant or the soil or other factors. However, it should not be forgotten that as a child requires special care in infant stage, so also a plant needs adequate care in the inchoate, *i.e.*, primary stage of its growth.

The power of resistance for most of the common fruit plants against adversities of soil and climate is obviously, much less at their primary stage of growth except for some species which are known to be hardy in nature. They increase their resistance and adaptability to the existing situation gradually till firmly established. The time required to get them well-established in the eco-environment varies and may be from less than a month to more than a year, depending on the plant species also other conditions. Evidently, necessary care is inevitable to them at this initial growth period of their life. Again, care should be taken with greater sophistication at still earlier part of their growth, *i.e.*, upto a few months after planting in most cases, keeping in view to meet the requirements to them as far as possible so as to ensure unobstructed and healthy growth.

The present Chapter has in fact, aimed at discussing those measures that are considered more essential to the fruit plants during the very primary stage of their growth, *i.e.*, a few months after planting of them.

18.2 Necessary Measures to be Taken

As a matter of fact, great care should be taken to the fruit plants during the first two years after they are planted in the orchard. Such measures are of great value to enable them to be more productive when they would be in grown up condition. A number of measures during the very early stage of their life would render benefit to them and among these, the following could be regarded to be of greater importance.

1. Provision of Desired Amount of Shade

It is essential that the newly planted propagating materials need to be saved from scorching heat, intense sunlight, heavy rain, hailstorm, severity of cold, fog, snow and such undesirable natural features. Protective measures against pilferers, cattle or other animals has also the necessity, if the orchard has not been provided with suitable protective device at its boundary as referred to under the Heading 6.4 in the Chapter 6. Therefore, such arrangements should be made around each propagating material as soon as it is planted or in some cases, even before planting. A number of methods are actually employed for such purposes in different parts of India, or are possible to be adopted. The more effective ones are enlisted below.

(i) Erecting a thatch around a plant is a very common practice in many parts of India. A thatch is made with broken branches of trees or bamboo slices with clippings of bushes or thorns *etc.* and dried leaves. Dried leaves of palmyra, wild date, coconut, arecanut, sugarcane tops *etc.* which are large in size offer good materials for the purpose. The method is not costly and satisfactory, if any special problem does not exist.

(ii) Sowing seeds of pigeon-pea in a circular manner around a planted material is also done by many orchardists especially to drive away off animals. But it hardly serves that purpose as leaves and twigs of most of the varieties of this plant are found to be eaten by animals. Nevertheless, benefit has been there to use this plant for shedding of leaves that add organic matter, addition of nitrogen to the soil besides that the soil erosion is also checked by these plants.

(iii) Planting of banana suckers around the newly planted material is found to be done in many parts of south India. The method has the merit in that, a high relative humidity is maintained in the inside but has also the demerit in that, banana being a heavy feeder takes up large amount of nutrients from the soil, for which the fruit plants are deprived.

(iv) Growing such plants around the propagating materials which are either not eaten by animals or which drive them away in having thorns such as *Ipomea carnea*, agave, cactus *etc.* is also a practice done by many.

(v) Provision of circular thatch made up of bamboo slices is a practice found very often in some parts of India. The method is no doubt, very satisfactory at low cost.

(vi) Erecting mud-wall around a plant is adopted by some orchardists, perhaps to protect against animals than natural adversities. Planting cacti and succulents over the walls is also done to ensure greater protection.

(vii) Making a roofing with polythene sheet in a slanting manner however, may be an effective material yet less costly. To erect this, as shown in Figure 18.1, two bamboo poles, each of about 1.2 metres should be infixed at one side of the plant such that these remain in one line and maintain same height above the ground. At the other side of these poles, two poles should be similarly infixed but the height of those above the ground should be lesser than those of the former two poles. Now, a polythene sheet is spread over so as to form a thatched roofing. Stretching the sheet to make it flat, the four ends of it should be tied to the four poles. On doing so, the polythene sheet will form as a slanting roof. Thus, it will protect the plants against heat, light *etc.* and rain water falling on it will also run out.

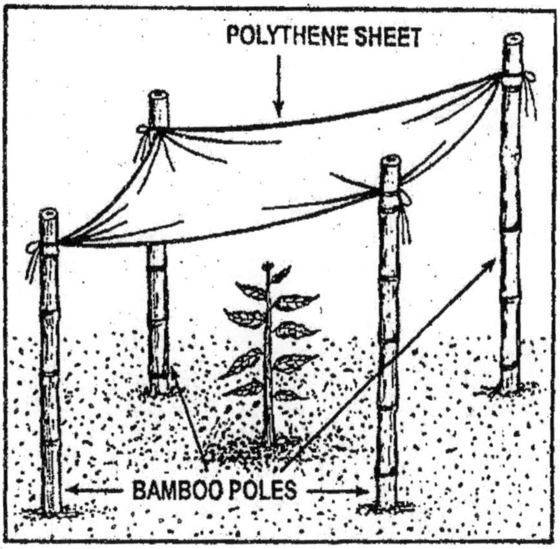

Figure 18.1: Slanting Cover with Polythene Sheet and Bamboo Poles.

2. Irrigation

Just after planting, moderate irrigation should be given to consolidate the soil, moistening of the roots and get them established in the soil withut delay. Thereafter, irrigation water should be applied with proper judgement, keeping in view the moisture status of the soil so that over or under irrigation is avoided. Regular irrigation as a routine practice may not in fact, become necessary if planting is done in rainy-season, unless there is a long spell of rainless period. But it should be a must, if planting is done in spring or other seasons.

When and how much irrigation water is to be applied depends on the requirement of the plants, moisture content of the soil and the texture of soil. In principle, the following two conditions should be borne in mind in irrigating the newly set plants, which are,

(i) Frequency should be more and

(ii) quantity should be less.

This is of particular importance where the soil is not of loose-texture. In general, it may be commented upon that light watering should be done everyday during the morning and the evening hours for the first 10-15 days after planting and thereafter, once a day upto a month. By that time, the plants get established and then irrigation on alternate days may be enough. In this way, the gap may be gradually widened such that one heavy irrigation is received by the plant once a week during the summer season.

As regard method of irrigation, making a small basin around a plant is very common in India. The basins are also linked with channels. The method is though suitable in loose-textured soil but where the soil is clayee, there is chance of water stagnation around the plants. Hence, irrigating the plants with buckets appears to be safer for any type of soil, at least during the first ten days of planting as this ensures application of the actual amount of water that is necessary. Where possible, water may also be let out through pipes which is more convenient.

The sprinkler or drip irrigation is no doubt, more befitting to irrigate plants at the early stage of growth and may be adopted, if affordable. If not affordable, a crude type of drip irrigation device as suggested by the present author (Ref. B.C. Mazumdar. 2004, and enlarged edition, 2022. *Orchard Irrigation and Soil Management Practices.* Daya Publishing House, Darya Ganj, New Delhi-110002) may be used.

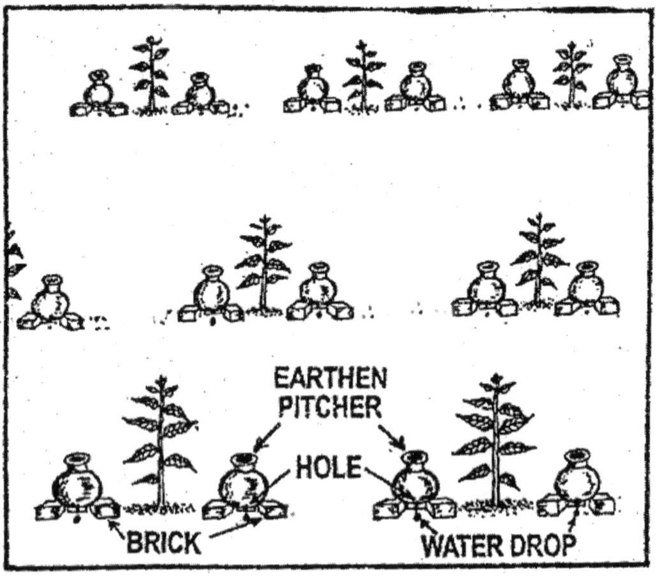

(a) (b) (c)

Figure 18.2: Crude Type of Drip Irrigation Device by using Earthen Pitchers. (Small Hole is made at the Bottom of each Pitcher. Pitcher is placed over Bricks, leaving the Hole Free).

To apply this device, as shown in Figure 18.2, earthen pitchers should be used. A small hole should be made at the bottom of each pitcher. Each pitcher should then be suitably placed on three of four bricks close to a plant in such a manner that the hole at the bottom is not blocked. It is better, if 2-4 such pitchers are placed at different sides of a plant, instead of one. The pitchers should be filled with clean water every day and that will drip down on the ground through the holes at the bottom. To avoid entry of birds inside the pitchers, they may be covered with wire-net pieces.

3. Drainage

For fruit plants in particular, drainage is *as important as irrigation is* and hence, should not be paid less importance. Stagnation of water around a plant may result due to heavy rain and more so, where the soil is of clayee nature. Making mud wall around a plant again worsens the condition. Where downward percolation is not adequate, draining out the excess water from around the base of the plants by making channels is no doubt, the step that is to be taken to save the plants against water-logging. In this connection, mention may be made to the fact that some orchardists heap soil around a plant with the idea that is does not come in direct contact with water. This is never desirable because heaping soil to the trunk of a plant renders great damage to the trunk and invites some diseases, notably, the *collar-rot* and other diseases of the trunk.

4. Protection of Graft Junction

Often it is noticed that when irrigation is done to the soil, irrigation water and even soil particles come to the grafted or budded part of the fruit plant. Water also comes to the grafted or budded junction when the plants are watered through cans, sprinklers or mugs.

It should be remembered that if water comes to the grafted or budded part, the union becomes weak and the scion easily loosens apart from the rootstock for which the scion dies. Entry of water in the cambium layer of the cut out parts of the scion and stock may also bring about fungal disease. To avoid water to come in contact with the grafted or budded region during irrigation, many horticultural scientists recommend tall grafting or tall budding, *i.e.*, to perform the grafting or budding operation at somewhat greater height on the rootstock. Wrapping the grafted or budded part with polythene over an extended part during irrigation of sprinkling of water may also prevent water to reach the operated part.

5. Application of Manures and Fertilizers

It should be remembered that although it is an admitted that after planting of the fruit plants in the orchard, nutrients are essentially required by the young plants for their unhampered growth and thereby, they require manures and fertilizers. But it should be remember that unintelligent application of manures and fertilizers would make great harm to the plants than any good.

It is advised by many horticulturists that no manure or fertilizer should be applied to the plants during the first 20-30 days after they are set in the orchard, in other words, until they produce new roots. Manure should be applied after that but not any fertilizer before 5-6 months of growth has taken place. Manures are organic substances and not so harmful. Besides this fact, they improve textural condition of the soil which makes increase in the water holding capacity of it. However, no manure should be applied to the soil unless it is well-decomposed. Well-rotten cowdung, dried form of farmyard manure, dried rural compost, filter-press cakes are some of the effective manures. Wood ash may be applied but in low does. Animal dung, sludge or poultry manure should not be used before 5-6 months of growth of the plants. *Vermicompost* as described under the Heading, 8.2 (4) in the Chapter 8 is an effective manure for application to the fruit plants at early stage of their growth.

Chemical fertilizers should be applied to the fruit plants after they have attained a growth of 5-6 months. In regard to fertilizer application at the early stage of growth of fruit plants, many horticultural scientists opine that *Straight fertilizers* which contain only one major nutrient, *e.g.*, urea are better to use than *Binary fertilizers* which contain two major nutrients, *e.g.*, potassium nitrate, or *Ternary fertilizers* which contain three primary nutrients, or *Complex* or *Complete fertilizers*. This is because, specific application of the nutrient could be made by application of Straight fertilizers.

Requirement of nitrogen is usually high during early stage of growth of the fruit plants to encourage vegetative growth in them and also for efficient utilization of phosphorus and potassium. On the other hand, sudden high uptake of nitrogen causes lodging of the plants, elongation of internodes, leathery leaves, high infection of diseases and other abnormalities. Hence, it is advisable to apply *slow releasing* nitrogen fertilizers, *e.g.*, urea formaldehyde (syn. Urea-methanal), which contains 35-42 per cent nitrogen, Oxamide [$(CONH_2)_2$], Neem coated urea, Sulphur coated urea, Super granulated urea *etc.* Phosphorus is required to stimulate root growth, new cell formation, leaf development *etc.* Potassium application is of great help to resist against diseases, insects and against climatic hazards.

Placement of fertilizers which indicates inserting fertilizers below soil surface by means of suitable implements give beneficial effect in some fruit plants when they have attained a growth of 1-2 years. Placement may be done close to the plants particularly in case of phosphate, or in band, or deep or as pellets and the like. Foliar application of the water solution of some fertilizers encourage high vegetative growth and stem maturity when applied to the fruit plants at early stages of their growth. Particular mention may be made to 0.5-1.0 per cent zinc sulphate solution which may be sprayed to the plants when they have made 8-10 months of growth. Absorption of zinc in the plant is rapid and it is highly mobile in the plant body as and gives rise to the growh-factor, auxin synthesis.

In respect of fertilizer application, close examination of the plants is necessary to check that *antagonistic effect* is not produced. It may be noted that some elements like nitrogen, phosphorus and potassium application in the plant may cause deficiency symptoms of copper, phosphorus application brings about deficiency of iron, zinc and copper, nitrogen, potassium and calcium application causes boron deficiency, excess of calcium causes phosphorus deficiency, calcium and magnesium cause potassium deficiency *etc.* Similarly, high intake of zinc causes copper and iron deficiency, molybdenum and copper cause iron deficiency and excess boron intake brings about molybdenum deficiency in the plant body. Amount of fertilizers should be applied in accordance with plant species but in general, it is advisable that the amount should be low though frequency of application may be more, if necessary.

6. Crop Protection Measures

Fruit plants are attacked by many insect or animal pests and diseases during early stage of their growth. They should be regularly checked and any attack by insect or animal pests or diseases, if observed should be controlled forthwith. Hand picking of insect larvae when observed and other physical methods of controlling pests and diseases should however, be given emphasis.

Chemical insecticides that are used are of various types. They may be of plant origin or animal origin. Among plant origin insecticides, some may be named as Nicotine, Azadirachtin, Nornicotin, Sabadilla, Rotenone, Anabasine *etc.* Among the animal origin inorganic compounds, some examples are Paris green, Lead arsenate, Sodium fluoride, Cryolite, Comp *etc.* Some examples of animal origin organic insecticides are Aldrin, Endrin, Endosulphan, DDT, BHC, Malathion, Parathion, Rogor, Carbaryl, Aldicarb, Lethane, Thanite *etc.*

Fungicides which are used to kill fungi come under varios types of chemical composition. For example, the sulphur containing ones are Zineb, Thiram *etc.*, copper containing compounds are Bordeaux mixture, copper oxychloride *etc.*, mercurial products are Phenyl mercury acetate, Methoxy ethyl *etc.*, heterocyclic nitrogenous compounds are Captan, Folpet *etc.* and the like. Which chemical pesticide should be used to control which insect pest or disease must be known thoroughly before it is used and should not be used arbitratily.

In the recent times, biological control of pests and diseases have drawn attention, particularly for their safety, Among the parasitoids or parasites, some of the agents that may be named as *Apanteles flavipes, A. subadinus, Leptomastix dectylopii,* larval parasite *Campoletis chloridae, Phytoselulus persimillis, Bacillus thuringiensis, Pasteuria penetrans, Metarhizium anisopliae,* nuclear polyhdrosis viruses *etc.*

It should be noted that inter-crop, if any grown in the orchard land during early stage of growth of the fruit plants may harbour some pests or diseases which attack the fruit plants. Accordingly, careful selection of inter-crops and their management are important considerations. It may be opined that *integrated pest management* deserves attention to protect the young fruit plants against possible attack by pests and diseases.

7. Control of Weeds

In early stage of growth of the fruit plants, much vacant space is left in the orchard and accordingly, weeds grow extensively. These unwanted plants, *i.e.,* angiosperms may be monocotyledons or dicotyledons. They are terrestrial or aquatic in their habitat and are annual, biennial or perennial in their duration of life.

Weeds are menacing problem in agriculture and horticulture. They take up large amount of water from the soil especially in non-irrigated areas and large amount of nutrients in irrigated areas and thereby, cause the crop plants to suffer. Some weeds are parasitic in nature and derive nutrients from the crop plants directly through special structure, known as *haustorium.* Many weeds harbour insect pests and disease organisms. Pollution of irrigation water and arresting drainage are also caused by them. For ability to survive in highly adverse edapho-climatic situation, production of huge number of viable seeds having easy germinability, many weeds multiply and grow very rapidly.

(a) Physical Methods to Control Weeds

Weeds must be controlled for the crop plants to grow successfully. There is a philosophical statement, *i.e.,* Mother Earth reminds the crop growers, " My land will not remain free of plants. Either you grow crop plants, or else, weed plants will occupy the land."

Hand-pulling of weed plants, *i.e.,* hand-weeding is the *eaiest, simple, ecologically safe* and *highly effective* method to eradicate weeds. This is done with suitable hoeing, mowing or tilling implements and sometimes by flooding. By hoeing, spade work or ploughing, stirring and aeration of the soil also take place which render benefit to the young fruit plants in the orchard. However, the method is slow, time-consuming with low output and costly for high wage paid to the labouters.

Growing extensive inter-crops in the vacant space during primary stage of growth of the fruit plants in the orchard is a highly effective and productive method to restrict or eradicate emergence and growth of weeds. But two important considerations must be kept in view in regard to inter-cropping. These are as follows.

(i) Judicious selection of the inter-crops is an important aspect as some species of inter-crops harbour many insect pests and diseases which may attack the young fruit plants also.

(ii) If irrigation and other agro-inputs that are provided to the inter-crops are not adequate, the young fruit plants will be seriously suffered for lack of availability of them.

(b) Use of Chemical Weedicides

Plentiful chemical weedicides have been on use now, among which *organic compounds* are found to be highly effective. Chemical weedicides (syn. Herbicides) are available in liquid or solid form and they are applied in three stages, which are, (i) pre-planting, (ii) pre-emergence and (iii) post-emergence, in accordance with weed and crop and are applied either to the soil or as foliar spray.

Chemical weedicides are classified according to their chemical groups and among many of them, few examples are stated below. (Ref. Sathe, T.V. 2003. *Agrochemicals and Pest Mangement*. Daya Publication House, Delhi-110035.

(i) Acetamides, which are mostly applied at pre-emergence stage. These include compounds, viz., Alachlor, Butachlor, Metolachlor, Dimethenamid, Propanil etc. (ii) Arsenical compounds which effectively control perennial weeds and examples are Monosodium methanearsenate, Disodium methane arsenate etc. (iii) Benzoics, such as Dicamba. (iv) Benzamides, such as Isoxaben. (v) Benzothiadiazoles, which include Bentazon. (vi) Bipyrilidiums, such as Paraquat and Diquat. (vii) Carbamates for example, Asulam and Phenmedipham. (viii) Cyclohexenones, which are active to control grasses, annual and perennial and examples are Clethodim, Tralkoxydim etc. (ix) Dinitroanilines, for example Trifluralin, Fluchloralin etc. (x) Diphenylethers, for example Acifluorfen. (xi) Imidazolinones for example, Imazaquin. (xii) Nitriles, such as Bromoxynil and Ioxynil. (xiii) Oxadiazoles, such as Oxadiazon. (xiv) Phenols which include Dinoseb. (xv) Phthalamates, for example, Naptalam. (xvi) Phyridines which include Dithiopyr. (xvii) Sulphonyl ureas which inhibit weed growth and causes necrosis, for example, Chlorimuron, Prosulfuron etc. (xviii) Thiocarbamates which produces toxicity to germinating weeds, for example, Diallate, Butulate etc. (xix) Uracils which inhibit root and shoot growth and include Terbacil, Bromacil etc. (xx) Ureas which cause laminar chlorosis and examples are Linuron, Diuron, Tebuthiuron etc. Which weedicide should be used in any particular condition needs to be intelligently decided before it is applied with consideration to the time of application. It may be stated in this connection that some species of weeds are easily destroyed only by foliar application of saline water. For example, *Parthenium* species which became serious in some parts of eastern India a few years before had much been controlled by foliar spraying with 15 per cent of common salt solution.

8. Training of Young Fruit Plants

It has been stated in the preceding Chapters that training is a horticultural operation which aims at *imparting any particular size, shape or form to a fruit plant*. To do the training operation, it may be necessary to cut out plant parts but it should be noted that the training operation may be performed without cutting any single plant part or not cutting even a single leaf.

As stated in Chapter 9 (Ref. 9.11-1-a), for erect-growing fruit trees or shrubs, one of the following three systems of training are followed. These are, (i) central leader system, (ii) open-centre (syn. Vase) system and (iii) modified central leader system. It should be noted that the above systems are not followed in a creeping plant like grapes, melons, strawberry *etc.* or in a low heighted plant like pineapple.

In regard to the training system to be followed, the following aspects should particularly be kept in view.

(a) Stage of Growth of the Fruit Plants

Training should be done to a fruit plant at *early stage of its growth*. This is because, in early stage, a fruit plant can be suitably bent to impart any particular shape. This will not be possible after few years of its growth, *i.e.*, when the branches will be thickened and hardened and hence, will be broken by the attempt of bending. However, cutting plant parts during the first few months after planting for the purpose of training is not desirable and the operation should be initiated when new growths have started in the fruit plant.

(b) Cutting Plant Parts in Installment

To perform training, cutting branches heavily at one season is not desirable as it may produce *great shock* to the young plants. Instead of that, it should be done lightly at two or three seasons.

(c) Training During Few Years of Growth

For most of the fruit plants, training is continued over few years.

In apple, to train by modified central leader system, the plants are de-topped at about 60 cm height during planting. During the first year, 3-4 scaffold branches are selected and these are headed-back to quarter, During second year, dormant pruning is done by cutting of the central leader and primary branches. Higher number of branches are retained on the trunk. In third year, thinning out and heading back of branches are done. During fourth year, growth of central leading shoot and sidewise branches are made to maintain proportional growth. At later stage of growth, the central leading shoot is headed back. For other temperate fruit trees also, training is continued upto four or five years of growth.

(d) Breaking Apical Dominance

In some fruit plants, *e.g.*, in guava, bael, aonla *etc.*, the physiological phenomenon, *apical dominance* is sometimes observed. That is, the apical part of the plant goes on increasing consistently without producing growth of side shoots. This happens due to higher accumulation of auxin at the apical bud or buds and the stem of the plant becomes thin. If this is observed, the apical bud(s) should be broken. By that, apical dominance is arrested and side shoots take up good growth.

(e) Bending of Branches

Fruit trees which are axillary bearers are usually found to produce fruits at the upper part of the shoots while the *lower parts remain unproductive*. In such fruit trees, *e.g.*, guava, peach, ber, aonla, fig *etc.*, *forced bending of branches by means of rope and hooks* make the entire shoots productive, *i.e.*, fruits are borne both at the upper part and at the lower part.

(f) Imparting Bushy Appearance

Training to assume bushy appearance increases productivity in some fruit trees, *e.g.*, pomegranate, custard apple *etc.*

(g) Avoiding Jungle of Growths

In citruses, ber *etc.*, sometimes many shoots appear closely together at one place. Regular *thinning out* of shoots where many growths have appeared in such trees should be a regular practice.

(h) Removal of Watersuckers

Regular emergence of *watersuckers at the base* appears to be a problem in many fruit trees. These growths do not give any production but draw much amount of water and nutrients from the soil for which the fruit trees are suffered. Hence, these unwanted growths should be cut off whenever found to appear.

(i) Sharp Cutting of Branches

Branches which are necessary to be cut should be done with a secaterur or a suitable knife to make cut surfaces flat to avoid contusion.

(j) Application of Fungicidal Waste

Whenever any branch of a fruit tree or shrub is cut off, the cut surface should be painted with a *fungicidal paste* to avoid its drying and also to disallow *opportunistic infection of disease organisms* through the cut surface.

9. Removal of Flowers

Sometimes infloresences are found to be produced in case of vegetatively propagated plants after few months of planting. This happens due to the development of flower-buds which were already present in the plants. It has been stated in the earlier part that these flowers do not produce desirable fruits but *hamper normal vegetative growth* of the plants. Hence, these should be nipped off whenever observed. However, this is not applicable for quick-growing fruit crops, *e.g.*, banana, papaya, strawberry, melons, cape-gooseberry *etc.* as they produce flowers early which give rise to normal fruits.

10. Protection against Heat and Light

Damage of plants due to heat or sunlight sometimes becomes a serious threat in tropical and subtropical climate. The destructive effect of heat is greater when it is accompanied by low relative humidity and high wind. Such condition results in loss of water both from soil and the plants and when it reaches to a high degree, the plants may be wilted. Apart from heat, intense sunlight also causes great harm to many plants among which, young litchi plants are classic examples.

To save the young plants against heat and light, the methods that have been stated above under the Heading (1), *i.e.*, provision of shade could be applicable. Wrapping the bark with paper may also be done to protect against heat and light. Making whitewash with lime on the trunk and main branches is also a good method to protect against sunlight as it reflects the light. To save the plants against storm, wind-break rows are no doubt, effective if suitably built up.

11. Protection against Cold

In temperate region and sometimes in subtropical situations also, cold poses to be a serious threat. Although some temperate fruit plants are very hardy in this respect and may withstand temperature much less than 0°C for some period, others plants are badly suffered due to heavy snow or frost. The damage due to low temperature may take place in the buds, leaves, wood, bark, cambium *etc.* apart from roots and evidently, young plants are more easily damaged than the mature trees. Again, nutritionally deficient and disease infected plants are more vulnerable to low temperature injury than the healthy and vigorously growing plants.

As regard protection of the plants against such extreme or very low temperature, many devices are there but most of them are costly. Heating obviously comes to the mind at first and this can be accomplished in various ways. Making fire by burning wood appears to be a simple method but this has many adverse effects to the plants and hence, to be done in a guarded manner if at all to be adopted. Undoubtedly, electrical heating is most effective but cost is very high. Exposure of infra-red ray at several places within 50 cm height from the ground level is also very effective, besides that electrical consumption is lesser in this method than to use electrical heaters.

Irrigating the orchard heavily is claimed to be beneficial in raising the soil temperature upto few degrees. This is believed to happen, as soil micro-organisms are activated in getting water and they *respire vigorously, generating heat energy*. Accordingly, inoculation of the soil at several parts with microbial fertilizer and then to irrigate the soil would also become effective.

(N.B. It should be noted that burning anything on the orchard floor or heating by electrical means kill plenty of micro-organisms and earthworms present in the soil which are of *great benefit* for crop production. Many years may be required for their re-appearance in the soil).

12. Mulching

To conserve soil moisture and control growth of weeds, the importance of mulching cannot be disregarded. If the entire orchard land is not possible to be mulched, the soil around each plant may be covered to derive the benefit.

13. Sanitation

Most of the fruit plants at primary stage of their growth are much susceptible to disease infection. Hence, adequate sanitary measures in the newly orchard is essential to get rid of the sources that are congenial substrates and hosts to the pathogens. Of particular mentioning is the farmyard manure or compost which should be used only when well-rotten. Making dumps of manure here and there in the orchard land is no doubt, a good measure for high incidence of disease infection to the fruit plants set and hence, must be avoided. Eaten parts of fruits or other materials dropped by birds on the orchard land, if any should be removed out in attempt to maintain cleanliness.

14. Close Observation

The orchardist should not in fact, forget to visit the newly set up orchard, if possible everyday and closely examine each plant, giving some time and take the action as felt necessary. As stated earlier, it is once again reminded to the orchardist the proverb prevalent in rural parts of West Bengal which states, "*The best manure of a crop plant is the dust of the feet of its owner. More frequently it gets, more quickly and virorously it grows.*"

15. Replacement of Plants in Mortality

Even with great care, some plants which have been set in the orchard may be found to be collapsed. These dead plants need to be removed and replaced by other healthy materials. As a matter of fact, the orchardist should keep this in mind beforehand. In accordance with, he should procure some extra plants from the nursery for the purpose and maintain them in the secondary nursery bed or in pots. Whenever there is death of any plant in the orchard, it should be replaced with a living one. As there may be difficulty to procure the same type of propagating materials again from the nursery later, it is advisable that the orchard owner purchase some extra plants along with those that are actually required and maintain them with care.

Chapter 19

Inter-cropping

19.1 Definition and Concept

Any crop plant grown in an orchard other than that of the fruit crop for which it has been built up, is known as an inter-crop

The inter-crops include those crops which are helpful in the following ways.

(i) Provided food to the human beings
(ii) Provided food to the domestic animals.
(iii) Produce something that fetch economic return to the orchard owner.
(iv Serve other purpose.
(v) Beneficial for the soil.
(vi) Beneficial for the fruit crop which brought about existence of the orchard.
(vii) Creates eco-environmental improvement of the locality.

In fact, inter-crop is a broad term, under which several types are recognized in accordance with (i) type, *i.e.*, species of crop grown, (ii) purpose of growing, (iii) duration of growing, (iv) method of culture adopted and so on. These are known by several terms, such as *filler, cover-crop, green manure crop* and *inter-crop* and are purported to carry the senses as stated below.[8]

1. Filler

As stated under the Chapter 14, a filler should always be a *fruit plant*. This is planted in the diagonal system of layout, *i.e.*, in the middlemost position of each

[8] Mazumdar, B.C. (2004). *Orchard Irrigation and Soil Mhanagement practices*. Daya Publishing New Delhi.

individual square in the square system, where two opposite diagonal lines intersect. The fruit plant used as a filler may be a tree, a shrub or a herb but that should not be of very low height like pineapple, strawberry, cape-gooseberry *etc.* According to the type of fruit plant, it is referred to as a filler tree, shrub or herb, or only as filler. Although one filler is usually set in the centre of a square or rectangle, sometimes two or even three fillers are also planted, if necessary.

2. Cover-crop

The crops which are grown in the orchard with the objective to protect the soil *against erosion* are termed as *cover-crops.* These crops therefore, act as a vegetation cover over the soil surface. The raising of cover crops is usually done in rainy-season when the danger of erosion is high and especially on a sloping terrain which is greatly threatened by erosion. An ideal cover-crop is one which has a highly ramified root system as may bind the soil firmly, slow-growing, bushy, needs less amount of moisture and nutrients from soil and does not harbour pests and diseases. Some examples of effective cover-crops are sunnhemp, dhaincha, lentil, cluster bean, berseem, lucerne, cowpea, moth, velvet bean *etc.*

3. Green Manuring Crop

These are leguminous or non-leguminous crop plants that are grown to add *bio-mass* (organic matter) or bio-mass and nitrogen to the soil. The crop is grown for a month or two and while the plants are still succulent, they are buried down in presence of adequate water in the soil, in order that they may get rotten and add organic matter. If the soil does not contain sufficient water, the plants are cut off and laid over the ground instead of allowing to grow further and become woody (lignified) which hampers rotting. Addition of nitrogen takes place from the nodules developed in the leguminous plants. Other aspects of green manure crops and green manuring have been dealt in under the Heading, 8.3 in the Chapter 8. It should be noted that some green manure crops also serve well as cover-crops.

4. Inter-crop

Inter-crop is in fact, an all-inclusive term denoting any crop plant other than the main fruit plant that is grown in the orchard. In that sense, filler, cover-crop and green-manuring crops come all under this group as has been referred to above. But in fruit culture, the term, inter-crop is more commonly used in a *specific sense, i.e.,* to refer to some particular type of fruit crops only. That is, the term is restricted to some short duration and economically important crops which usually provide food or other necessities to man. These are grown as seasonal crops, following some particular crop-rotational programmes and evidently, with the objective to secure some supplementary produce from the orchard land.

In the present Chapter, the term, inter-crop has been used and interpreted as it is implied in a specific sense only and not as a general term.

19.2 Plants Grown as Inter-crops

The inter-crops comprise those crops which have *economic benefit* to the human beings either as food or for other purpose. The crops that are most commonly grown include various types of vegetables, condiments, spices, pulses, oilseeds, fodder crops, fibre crops, narcotic crops, medicinal and aromatic crops, floricultural crops, some cereals and millets and even some short duration and low heighted fruit crops like pineapple, strawberry, cape-gooseberry *etc.*

It may be noted that the inter-crops that are to be grown in the early life of the orchard may not be the same as those which would be grown when the fruit trees become mature. As the present Chapter aims at discussing inter-cropping that may be done within first few months or years after the fruit plants have been set, it may be stated that at this very early life of the orchard, inter-cropping could in fact, be done *extensively and selecting a wide variety of crop plants.* Such an opportunity exists for two reasons. These are firstly, adequate vacant space is left between the fruit plants and secondly, fruit plants being very young would not provide shade over the inter-crops. But although such possibility is there, the orchard owner should heedfully consider certain aspects as regard choice of the inter-crops to be grown. These may include demand of the produce in the market, distance and accessibility to the market, storage facility in the nearness, industrial requirement if any, requirement of inputs by the inter-crops and their availability, cultural procedure to be followed *etc.* Whether any such crop has any good or bad effect over the fruit plants that have been set in the orchard is important factor that needs to be adequately considered.

19.3 Advantages

In a newly set up orchard, inter-cropping offers a number of advantages and the important ones may be enlisted as follows.

(i) Productive Utilization of Unused Land

The fruit plants are set in the orchard maintaining the required planting distance. As stated under the Chapter 13, the planting distance of the fruit plants is determined keeping in view the canopy spread of the plants that would be assumed when they would be in fully grown up condition. Now, when the fruit plants are in primary stage of growth they occupy very little space but the planting distance that has been maintained remains the same. That is to say, wide space is left in the orchard between the plants when they are in so young stage.

Instead of leaving this spacious area of the orchard only as a fallow land, it is a logical approach to make utilization of that by raising some suitable crop(s). This justifies inter-cropping in the new orchard, the objective of which is to make full utilization of the land in the orchard in a productive manner.

(ii) Source of Income

Until the fruit plants that are set in the orchard attain bearing maturity, they produce commercially important fruits, there is no income from the orchard. On the other hand, during this (gestation) period, regular expenditure becomes necessary to maintain and induce growth of the plants by applying necessary agro-inputs. Therefore, it is warrantable if any extra income is possible to be made from the orchard during this phase. The inter-crops are undoubtedly, effective means for this purpose as the orchard owner can make a substantial earning by judicious selection and culture of these crops.

(iii) Weed Control

It should not be forgotten that a cultivable land never remains barren, meaning thereby, that if crop is not grown in it, weeds will grow. In fact, the large part of the orchard land left between the plants becomes an ideal place for profuse growth of weeds. These unwanted plants withdraw large amount of moisture and nutrients from the soil and harbour many pests and diseases. This menacing problem can evidently be averted by growing inter-crops in the orchard.

(iv) Benefit to the Fruit Plants

When inter-crops are grown intensively by applying adequate inputs, the fruit plants are also sometimes benefited indirectly from the applied inputs.

(v) Increased Relative Humidity

By growing inter-crops with copious irrigation water, the relative humidity in the orchard are a is increased for which irrigation requirement to the fruit plants becomes lesser in higher relative humid atmosphere.

(vi) Improving Subsoil

The roots of the inter-crops after harvest are decomposed and thereby, improve textural condition and fertility of the soil. Besides, if pulses are grown as inter-crops, sufficient nitrogen is also added by the nodules attached to their roots.

(vii) Checking Soil Erosion

Some of the inter-crops act as effective cover crops as well. That is, their roots highly proliferate and bind the soil for which erosional loss of soil is much retarded.

19.4 Disadvantage

Despite other advantages, some drawbacks may also be there in inter-cropping. This usually happens owing to the fact that attention of the orchard owner becomes greater to the inter-crops than the fruit plants. It is often noticed that the inter-crops receive adequate manure, fertilizers and culture but fruit plants are not provided or timely supply does not take place in them.

Some of the inter-crops *withdraw heavily* from the soil for which the fruit plants are deprived. This particularly results if the inter-crops are not provided with adequate amount of manure and fertilizers. Similarly, if adequate crop protection measures are not made available to the inter-crops, they are affected by a number of pests and diseases. Due to harbouring of these agents, the fruit plants may also be attacked.

19.5 Conclusion

To settle the moot point, *i.e.*, whether inter-crops should be grown in the early life of the orchard to derive the benefits or dispensed with for any harmful effect rendered to the fruit plants or the orchard soil, the orchard owner should call to mind the basic fact, *i.e.*, the orchard has been established for those fruit plants which have been set in it and not for the inter-crops. Thus, the fruit plants are the *primary materials in the orchard and the inter-crops are auxiliary*, in other words, supplementary crops only. In clear terms, the fruit plants set are permanent members in the orchard while the inter-crops are grown as temporary crops. Having considered this basic concept, what is necessary for the orchard owner is to pay much greater attention to the fruit plants than the inter-crops as the *orchard has been established for the former plants.* In accordance with, selection and culture of the inter-crops are to be done.

In the selection of the inter-crops, care should be taken on a number of aspects, among which the following two are of greater relevance.

(i) Such an inter-crop should not be chosen to grow which may cause harm to the fruit plants or to the orchard soil condition. Thus, those crops which withdraw very high amount of nutrients and moisture from the soil should as far as possible be avoided. This is because, the soil gets deficient and the fruit plants may be deprived sooner or later. (It is pointed out in this context that high application of inputs is although theoretically considered in their culture, in many a case, this is not done so in practice.)

Those crops that are more vulnerable to the attack of those pests or diseases which may attack the fruit plants should be avoided. Example may be cited as of viruses in particular. Fruit plants are very often attacked by viruses from the inter-crops which have already been infected. Wilting of the fruit plants due to the soil-borne fungi belonged to the genus, *Fusarium* also may result due to improper selection of inter-crops like *Cajanas cajan* or such crops that are much prone to *Fusarium* attack. The insect pests, mealy bug and the bark-eating caterpillar, *Indarbela tetranois* are found to spread quickly from some particular type of crops. Mites and nematodes could also be cited as other examples. Evidently, such crops should not be introduced into the orchard only to invite hazards.

(ii) As far as possible, such crops should be selected which render benefit to the fruit plants, or the soil. Undoubtedly, the leguminous crops are very good for this purpose for the reason that their roots add sufficient amount of nitrogen to the soil when cropped.

Where there is danger of soil erosion, such inter-crops which also act as good cover crops should be selected. Some of the *leguminous crops* are known to be very effective in this respect. Apart from yielding pulse grains, they add adequate nitrogen and organic matter to the soil and protect the soil against erosion and leaching.

Besides selection, culture of the inter-crops in the young orchard also needs adequate attention. A number of measures should be judiciously considered in this regard, among which the following are of greater significance:

(i) No inter-crop is to be grown close to any fruit plant and at least a gap of 2.5 metres should be maintained between a fruit plant and an inter-crop when inter-cropping is desired during the first few months after planting of the propagating materials.

(ii) Inter-crops should not be grown haphazardly here and there in the orchard land. Instead of that, *separate blocks* should be made between rows of fruit plants and cropping should be done in those blocks.

(iii) Irrigation and drainage channels should be made separately for the inter-crops. In other words, the same channels intended for the fruit plants should not be used for the inter-crops.

(iv) Amount of manure, fertilizers and irrigation that are necessary for the inter-crops need to be applied without niggardiness. On the other hand, somewhat higher application is advisable so as to leave a residue.

(v) Adoption of crop protection measures is a very important necessity in inter-cropping. Apart from controlling pests and diseases when appear, preventive measures should be adequately taken and prophylactic spray of plant protection chemicals is also to be considered.

(vi) Regular weeding is an essential step in inter-cropping. When weeding of the inter-crop is done, weeding around the fruit plants should also be done simultaneously.

(vii) After harvest of the inter-crops, the stubbles should be removed, the block of land should be ploughed or spaded and levelled.

As a matter of fact, inter-cropping in an orchard especially when it is young should be done intelligently. If this is carried out in a judicious manner, not only that an extra production and earning is secured from the orchard, but it also renders adequate benefit to the fruit plants, *i.e.*, the primary crop, directly or indirectly. If inter-cropping is not practised in proper way, harm will be rendered to the soil and productiveness of the fruit plants may be seriously hampered, either quantitatively or qualitatively or both.

Chapter 20

Set-up for Supplemental Income

20.1 Lack of Adequate Return

Sometimes it comes to notice that adequate profit is hardly obtained from a commercial scale orchard. It may be said that running into loss is even not uncommon in case of many orchards in India. For such undesirable happening, a section of fruit growers in the country now prefers to utilize their orchard land for other purpose also instead of restricting to the culture of fruit crops.

There are in fact, a number of factors that could be said to be responsible for inadequate or no return secured by the fruit growers from their orchards. These are associated with either production pattern of the fruit crops grown in the orchard or in their disposal system or both. Among a host of factors, some that draw much attention may be stated as follows.

(i) Taking the case of mango which is the most important fruit crop in India and occupies an area of about 2400 thousand hectares is known to be threatened by the problem of *bienniality* or *irregularity of bearing* in case of many varieties. This accounts for sub-optimal or no production of fruits in some years and thus, an economic loss is incurred by the growers of this fruit crop. Malformation of inflorescences or vegetative shoots is another problem that hampers productivity of mango trees.

(ii) Attack by *pests and diseases* is a serious bottle-neck that brings down the yield performance of many fruits. Some problems like *wilt* disease in guava, *decline* in citruses, *viruses* in papaya, *anthracnose* in grapes, *die-back* in some fruits, *San jose scale* in apple, *leaf-curl* in peach, *bark-eating caterpillars* in some fruit trees *etc.* tend to depress growers for production owing to the possible risks associated with cropping of such fruits.

(iii) Seasonal availability of fruits in short duration is a factor of low return as compared to many field crops which are available over a relatively long period.

(iv) Due to *inadequate cold storage facilities* specifically for fruit crops belonging to different species, the excess production could not be stored for disposal at later time to fetch higher return.

(v) Increased cost of production consequential to high cost of inputs and labour cost make the growers boggled to go for higher production of fruits by making increase in area.

(vi) In a healthy marketing structure, not more than 20 per cent of the profit from the harvest of the produce till it reaches the hands of the consumers should go to the middlemen and 80 per cent should be shared by the growers. But in most cases, the profit earned by the *brokers intermediaries* is much higher for which the growers are deprived.

20.2 Steps to Increase Return by Supplementary Earning

It must be admitted that fruit growing needs emphasis to make increase in fruit production in India and in this endeavour, adequate earning from orcharding also needs to be ensured. Fruit growers in different zones of the country should form co-operative societies and take suitable steps to secure higher return from this horticultural avocation and in this drive, participation of the government agencies has also the necessity.

Mention may be made in this connection that some orchards could however, be utilized to generate income from activities apart from production of fruits. There may be many such pursuits to earn *supplementary income from the orchards* which could be taken and run successfully without hampering fruit production in the orchard.

It should be noted that if any such project is planned to be taken up, space for setting up of it should be allocated in the orchard at the time of its establishment. Suitable infrastructural arrangements are also necessary to perform while the orchard is established.

A few projects that could be launched in an orchard might be brought to notice as follows.

1. Plant-Nursery

Preparation of propagating materials including seeds of various types of plants and their selling would be the activity of the nursery. However, the nursery may also sell such type of materials on procuring those from other sources. The nursery if installed, should preferably be located near one of the boundaries of the orchard.

2. Selling of Materials Used in Garden

The orchard may earn some profit by opening a counter for selling of garden implements, watering cans, plant nutrient salts, plants growth regulating chemicals, pesticides *etc.* as well as books on culture of horticultural plants. The counter should preferably be opened near the office room.

3. Poultry

Well-managed poultry farm in the orchard has justification for adequate extra earning. If necessary, the eggs may be sent to distant places and if it is desired to do so, they should be dipped in lime water and dried so as to avoid rotting within few days. The excreta of the poultry birds may suitably be composted for use as manure. Poultry should be located near a boundary of the orchard. (If the orchard is fairly large in size, rearing of goats and sheep may also be considered by the orchardist).

4. Beekeeping

Production of honey by beekeeping in the orchard needs encouragement for extra earning. The insects would be of much help for pollination of the flowers of many fruit plants also.

5. Mushroom Growing

The orchard may have considerable earning by growing mushrooms. Culture of oyster and paddy straw mushrooms could be done easily. Although growing of white button mushroom requires efforts but they have high demand in the market.

6. Fruit processing and preservation

High return may be obtained by drying, processing and preservation of fruits, besides the fact that the fruits which would be left undisposed could be utilized. If large-scale processing is difficult, small-scale unit may be set up in the orchard. But adequate care should be taken that the smoke which if emanates from the processing centre does not come in contact with the fruit trees. If installation of processing and preserving centre is to be done, it should be located close to any boundary.

7. Hiring for Picnic

Nowadays, many orchardists find interest to hire the orchard land for the purpose of picnicking. Although the orchard may earn income from this hiring but adequate care is necessary to ensure that fruit trees in the orchard are not damaged. It may be pointed out that the smoke emanated by the fire in cooking food may cause harm to the fruit trees if comes in contact to them. Food packets left by the picnickers may also hamper sanitation of the orchard. Hence, if the orchard land is hired for the use of picnic, it could be done at the boundary and the allocated land should be suitably fenced for this purpose.

8. Building up Resort

Orchardists in some parts of India have built up small houses inside the orchard for use as *resorts*, naming as eco-friendly or garden resorts intended for short stay of the visitors. If such a project is desired to be set up, it (those) should be allocated near the boundary of the orchard.

9. Setting up Children's School

Many orchardists in India have set up or are planning to establish children's school, kindergarten or crèche inside the orchard. Two to three well-built rooms with roofs, few benches, chairs, tables, books, toy-horses, toy carts, other toys *etc.* may be needed to maintain. Robust branches of trees may be suitably utilized to make swings however, with care. Such schools within the orchard would have great effect on mental behavior of the children, if they are not come down on for small fault. The orchardist himself may run such a school or turst other person(s) as hireling and earn.

Chapter 21

Maintaining Maps and Records

21.1 Principles

Particulars on where and how the orchard has been established, what are the components of it, what materials are under the possession of the orchard and other details are no doubt, within the knowledge of the orchardist. Similarly, during running condition of the orchard, what activities are carried on it inputs applied, expenditure incurred *etc.* are also known to him. However, it may be difficult for him to keep in mind all such informations. But it may be mentioned that knowledge of these particulars has the necessity to the orchardist *always for efficient running of his orchard.* Such information may also be of great help to those who would be interested in orcharding.

Therefore, it is essential that all activities should be recorded in *pen and paper* and this aspect should be given much importance. As many informations as possible should be recorded and separate registers, folders, files *etc.* should be maintained to record informations of different items. Recording materials should be of better quality and of long durability. The orchardist should feel free to show those to a person having interest on farming of fruit crops. He should also satisfactorily answer to the questions if asked by interested persons.

21.2 Particulars to be Maintained after Establishment

1. Surveyed Map

Survey of the orchard area and preparing suitable map to the scale allocating the different components, cropping plots, accessories *etc.*, are no doubt, the step that should be taken up in the first place. Apart from a complete map for the entire orchard, it also becomes necessary to prepare detailed maps for separate blocks or parts of the orchard.

Several copies of the maps should be made and carefully maintained. The copies of the maps should also be displayed at the orchard office to make noticeable to the orchard staff and visitors. It is also advisable to arrange for making display of the maps of the different blocks at the respective guide-posts installed at different locations of the orchard as stated under the Chapters 10 and 11.

2. Record of Cropping Plots

As stated under the Heading, 13.2 in Chapter 13, particulars of the cropping plots should be maintained in *separate registers* after those have been laid out in the orchard.

3. Record of Fruit Plants

The detailed informations of the fruit plants set in each cropping plot should no doubt, be recorded carefully in a separate register.

The particulars that are essential to be kept in record are botanical species of the fruit plants, horticultural variety *etc.*, source of procurement, price, method of propagation followed, rootstock used for grafted materials, age of the materials at the time of planting, date of planting *etc.* If there is mortality of any plant after it is planted, record of that should also be made, specifying the period after which it has taken place, possible cause if known, whether the same has been replaced by other material and the details of that.

4. Inventory

It is essential that a list of articles, both movable and immovable that have been procured for the orchard should be maintained and for this purpose, a separate stock-book is necessary in which the details of each article should be recorded. Among other particulars, the name of each article, its specifications, source and date of procurement, price *etc.* must be recorded. The stock-book should also accompany a folder to preserve the cash-memo, warranty papers, if any and operating instructions for each of the article. With use, if any article is damaged or dead or written-off or auctioned, the details should no doubt, be kept in record. Further, the stock-book, the folders and files should not miss in recording the details of any article, if purchased later during the running condition of the orchard.

21.3 Particulars to be Maintained during Running Condition of the Orchard

A number of records are necessary to be maintained during running condition of the orchard. It should be noted that as many informations as possible should be recorded as each of these informations proves to be helpful some day or other. Nevertheless, recording of the following informations appear to have greater necessity.

(i) Growth rate of the fruit plants.
(ii) Tillage operations performed.
(iii) Manures, fertilizers, growth regulating chemicals *etc.*, that have been applied.
(iv) Irrigation applied.
(v) Training of the young plants.
(vi) Pruning of the mature trees done.
(vii) Crop-protection measures adopted.
(viii) Yield performance.
(ix) Quality of the produce.
(x) Inter-cropping, if any is practised.
(xi) Disposal of the produce.
(xii) Labourers employed.
(xiii) Machineries employed.
(xiv) Problem faced, if any.
(xv) Expenditure and return and so on.

In what manner the informations are to be recorded may be known from an expert in fruit culture and also, from an experienced accountant. As a matter of fact, the orchardist may take advice from others but what should be the most convenient method should however, be decided by himself. In course of time, it may also become necessary to modify the methods within certain limits to suit the conditions.

Maintaining records of a fruit farming industry should in fact, be done in consideration to many aspects. Some of the basic principles that are to be kept in view for this purpose may be stated as follows:

(i) Records of the given orchard and its activities should be done in such a manner that those become easily understandable not only to the record-keeper himself but also to any other person having some background on fruit culture and even without spending much time.
(ii) Entries and records should be done in simple form as much as possible without making complications.
(iii) Unnecessary writings should always be avoided.
(iv) Separate books, folders and files should be maintained, instead of jumbling up all in one or two.
(v) Delay in recording should never be indulged and it should be a routine practice to keep records regularly.
(vi) For greater convenience, any modification of record-keeping, if felt necessary may be done.

(vii) The files, folder,s papers *etc.* should be of better quality such that they could be stored well for many years.

(viii The mode of recording should be known to a number of persons and not one only, for in absence of one person, another person should be able to carry on the job.

(ix) Records and papers should be maintained in suitable almirahs under lock and key and at times, treatment with insecticides should be able to even become necessary at that place for their safe preservation.

(x) Taking the help of a computer is definitely of enormous advantage in this job.

Chapter 22

Concluding Part

India has made a great stride in fruit production at the present time. The country now produces about 100 million metric tonnes of fruits annually from an area of about 6.5 million hectares. But it needs pointing out that the area under fruit crops is only 4 per cent of the total area under agriculture in the country. Evidently, the area is needed to be increased to meet the requirement of the ever increasing population.

It is a general observation that many fruit growers in India do not find much interest to increase area under fruit crops fearing inadequate profit. However, profit could be made significantly high on attention of the following aspects in particular.

1. Judicious Cropping

Risk factors involved in fruit farming could in general, be said to be greater than growing other crops. Some of the more common factors that are associated with less satisfactory return from the orchards could be stated as follows.

Gestation period in case of majority of the fruit crops is much higher than the commonly grown field crops and accordingly, most of them necessitate several years of growth to produce fruits of commercially accepted stage. Thus, there is no income from the orchard land during the *first few years of planting*. Barring nuts, most of the fruits being soft textured are easily perishable under ordinary atmospheric storage. Although they can be stored for some period at low temperature but the temperature requirement in cold storage differs according to species and varieties of fruits. In fact, installation of cold storage houses separately for different types of fruits has practical difficulty. Fruits belonging to different species give rise to produce for a short period in the year and failure of their timely disposal incurs great economic loss. In regard to disposal of the harvested fruits, the *share of the middlemen* (intermediaries) should although not go for 20 per cent of the total profit, in reality, their share goes to much higher level in most cases.

However, apart from earning profit from commercial fruit culture, consistent increase in the production of human consumable fruits cannot be looked down upon in India to subserve as *protective food to the teeming millions*, running of many industries, rearing of domestic animals, eco-environmental improvement and for other purposes.

Earning profit from an orchard has beyond doubt, the necessity particularly when it has been set up for commercial purpose. Several steps are involved to minimize loss from an orchard and turn it to a profiteering investment. Among other steps, the following cropping procedure may deserve adequate attention in an orchard.

(i) Choice of the species of fruit crops is no doubt, the most important factor. For example, in Allahabad-Banaras area in U.P., the fruit growers had a tendency to grow more and more of guava and even in few decades ago, many new orchards were found to come up with guava as the crop. Eventually, the supply of the fruit had been much higher at that time as compared to the demand. Similar problem is faced by many viticulturists in some parts of Maharashtra where grape is abundantly grown. Under such condition, the fruit growers have to think intelligently the farming of fruit crops belonging to different species instead of restricting to the traditionally grown species of fruit crop only. In some cases however, there is difficulty to set up new plantations for raising fruit crops of other species. Thus, mention may be made in this connection to the viticultural practice in. Maharashtra. The grape growers in this tract have already spent huge amount of money to build up permanent framework to train their grapevines and switching over to grow other fruit crops leaving behind the permanent framework does not appear to be a desirable and economic proposition.

(ii) Indian fruit growers have a tendency to grow in large-scale the same varieties of fruit crops(s) year after year, belonging to a species. For example, in mango growing zone in north India, the Langra, Dusheri and Chowsa varieties are extensively grown. Similarly, in south India, Mulgoa, Totapuri, Neelum *etc.*, in Andhra Pradesh, Bananpallee and Suvarnarekha, in western India, Kesar, Rajapuri, Dalimbya *etc.* and in Bihar, Zardalu, Malda and Bathua varieties occupy large area. In the famous Malda-Murshidabad region of West Bengal in growing mango, Himsagar, Bhuto Bombai, Shahdulla, Fazli *etc.* are much grown among plentiful delicious varieties that are adaptable there. In case of banana, in south India, the largely grown varieties are Poovan and Rasthali, in Maharashtra, the Basrai dwarf and in Andhra Pradesh, Tella Chakrakelli variety.

In Maharashtra, Nagpur Santra is the principal variety of mandarin orange while in Assam, the Khasi variety and in south India, the coorg variety. In guava, the abundantly grown varieties in eastern U.P. are Safeda, Chittidar and L-49, in Maharashtra, Nasik and in West Bengal, the Baruipur variety. In papaya, in south India, the Coorg Honey Dew, Pusa and the Coimbatore and northern India, the Washington and in east India, Ranchi variety are largely grown. In grape, the varieties, Tas-e-Ganesh, Sonaka and Chaman in Maharastra, while in Punjab, Perlette, Pusa seedless, Arkavati, Arkakandhan varieties are largely grown. In sapota, Kalipatti and Chhatri are much grown in Maharashta and in pomegranate, the Alandi and Dholka varieties get emphasis in western India. Similarly, in apple, the Ambri, Blood red, Baldwin *etc.*, in Kashmir, Red and Golden delicious in Himachal Pradesh and Rome beauty in the Nilgiri hills get high preference for cropping.

It may be brought to notice that growing same varieties of a species of a fruit crop in a region might be a factor that *adversely affect the balance of demand and supply*. Growing several varieties of a species of fruit crop in a fruit growing zone appears to be a justifiable approach for higher economic return.

(iii) Timely application of inputs has a great impact in quantitative and qualitative productiveness of a fruit crop. It should be noted that application of organic manure in rainy-season and during development of fruits is far from justification. In order to control the insect pest, mango hopper, spraying mango trees with insecticide after the flowers have opened, seriously hamper pollination and fruit set. Leaves that naturally fall from the trees in an orchard greatly help in *mulching* the soil and add *organic matter*. Hence, their removal from the orchard floor is not justifiable.

(iv) Formation of *co-operative societies and their efficient functioning* have great economic benefit in fruit culture. Heavy implements, sprayers, dusters could be purchased by these societies for use of the members who are associated and also to others on rental basis.

2. Inter-cropping

Properly done inter-cropping in the orchard as referred to under the Chapter 19 could be said to fetch high return from the orchard land and thereby, avert economic loss. It may be stated in this connection that fruit growers may secure high profit by growing some inter-crops, notaby, as medicinal and aromatic plants.

3. Organic Cropping

Of late, *organically grown fruits* are known to have high demand and are sold at high price. The orchardists should pay heedful attention to this aspect. Growing at least some fruit crops in organic way is likely to provide high economic return.

4. Growing Minor Fruit Crops

Establishment of orchards in physico-chemically low-graded land and uneven physiography might be an effective step to secure *high income in many cases*. The land to set up fruit plantation could be procured easily and at relatively low price as the demand for such type of land is much less for growing field crops.

The species of fruit crops which are otherwise categorized as *less-known, less-utilized, under-utilized* or *minor fruit crops* could be successfully grown in land of such type (Ref. B.C. Mazumdar, 2004, 2022. *Minor Fruit Crops of India*: *Tropical and Subtropical*. Enlarged Edition. Daya Publishing House, Delhi 110002). Fruits under such group grow and produce satisfactorily with less care and use of inputs and hence, production cost of these fruits is much less. However, the market price as fresh fruit is appreciably high in fruits of many of these species.

It may be brought to light in this connection that fruits belonging to many species of these fruits offer as excellent materials in processing and preservation. Hence, if such fruit crops are planted, the fruit growers themselves can produce the products. Some of the processed products that could be made from the minor fruits are stated below.

- ☆ Jam: Jamun, karonda, aonla, jackfruit, tamarind, custard apple, bael, ber, mulberry, cape-gooseberry, wood apple *etc.*
- ☆ Jelly: tamarind, jamun, karonda, Barbados cherry, wild guavas *etc.*
- ☆ Preserve: ber, aonla, bael, ker, sangari, karonda *etc.*
- ☆ Candy: aonla, ber, karonda, wild-date, tamarind *etc.*
- ☆ Glazed fruits: tamarind, ananas, aonla *etc.*
- ☆ Juice/Beverage/Squash: tamarind, bael, jamun, karonda, phalsa, ber, aonla, mulberry, wood apple *etc.*
- ☆ Wine: madhuca, wild apricot, wild date juice, palmyra juice, karonda, ber, Indian fig *etc.* (N.B. it needs mentioning that the fruit growers themselves should not resort to making wine. Because, if the juice is fermented in a wrong way, the wine may be poisonous).
- ☆ Chuteny: karonda, tamarind, aonla, wood apple *etc.*
- ☆ Sauce: karonda, tamarind wood apple *etc.*
- ☆ Pickle: aonla, karonda, sangari, ber, tamarind *etc.*
- ☆ Dehydration: bael, jackfruit, karonda, phalsa, ber, sangrari, mulberry, date palm, lasoda, tamarind, aonla *etc.*
- ☆ Canning: ber, star-apple, aonla, jamun *etc.*
- ☆ Confectionary: hog-pulm, aonla, tamarind *etc.*

Apart from using minor fruits, processing and preservation of major fruits could also be done by the fruit growers themselves. Even prematurely dropped fruits of mango and other fruits of mango and other fruits could also be utilized in drying and making other preserved products. It can be commented upon that formation of co-operative societies amongst the fruit growers might be of great help in regard to processing and preservation of fruits.

5. Leasing Out Orchard

Where there is difficulty in management of an orchard or in disposal of the fruits harvested from it, the orchardist may consider of leasing out the orchard to any person or agency. Under Indian condition, such a practice is prevalent in many areas. Leasing out is usually done to the lessee when the fruits in the trees have grown to some extent. Lessee enters into agreement with the orchardist on examining the orchard trees and pay the latter accordingly and partly in advance. The orchard is protected by the person who has taken lease against pilferage and sell the produce either directly or through agent.

Appendix I

Botanical Names and Families of Fruits Mentioned in the Text

Almond (= Bilati badam, Badam vittulu) : *Prunus amygdalus* Bats. (Fam. Rosaceae)

Aonla (= Emblica, Amlai, Myrobalan, Indian gooseberry, Nelli) : *Phyllanthus emblica* L. (Fam. Euphorbiaceae)

Apple (= Seo, Sebu, Apel Chunt): *Malus pumila* Mill.) : (Fam. Rosaceae)

Apricot (= Khubani, Chohara, Zardalu, Khurmani): *Prunus armeniaca* L. (Fam. Rosaceae)

Bael (= Bengal quince, Bilva, Bel, Maredu): *Aegle marmelos* Corr. (Fam. Rutaceae)

Banana (= Kala, Rambha, Kela, Arati, Vazhai) : *Musa paradisiacal* L. (Fam. Musaceae)

Barbados cherry (= Barbados cherry) : *Malpighia glabra* L. (Fam. Malpighiaceae)

Ber (= Kul, Topakul, Bor) : *Ziziphus mauritiana* Lamk. (Fam. Rhamnaceae)

Blue berries (= Sad berries) : *Vaccinium* (Section: Cyonococcus) species (Fam. Ericaeae)

Calamondin (= Hazara, Chinese orange) : *Citrus madurensis* Lour. (Fam. Rutaceae)

Cape-gooseberry (= Tepari, Tipari, Mokowi) : *Physalis peruviana* L. (Fam. Solanaceae)

Carambola (= Kamrakh, Kamranga) : *Averrhoa carambola* L. (Fam. Oxalidaceae)

Cashew (= Kaju, Mindiri, Geru bija, Munthamamidi, Hijli badam) : *Anacardium occidentale* L. (Fam. Anacardiaceae)

Citron (= Mahalunga, Gilam, Turnanj, Rusakam, Lungamu): *Citrus medica* L. (Fam. Arecaceae)

Coconut (= Nariel, Narkel, Daab) : *Cocos nucifera* L. (Fam. Rutaceae)

Cordia (= Cordia) : *Cordia dichotoma* G. Forest (Fam. Boranginaceae)

Currant (= Red flower currant, Blood currant) : *Ribes sanguineum* Pursh (Fam. Grossulariaceae)

Custard apple (= Sweet sop, Sugar apple, Ata, Sharifa, Sitaphal) : *Anona squamosa* L. Pursh (Fam. Annonaceae)

Date (= Khejur, Khajur, Kukyan, Karmah) : *Phoenix dactylifera* L. (Fam. Palmae)

Dillenia (= Chalta, Karmal, Carambel) : *Delenia indica* L. (Fam. Dilleniaceae)

Diospyros (= Gaab, Tumbi, Kalatendu Kendu) : *Diospyros perregrina* Gaertn. (Fam. Ebenaceae)

Durian (= Civet) : *Durio zibethinus* L. (Fam. Bombaceaceae)

Elephant apple (= Wood apple, Kaith bael, Vilanga, Kapith) : *Feronia limonia* (L.) Swin. (Fam. Myrtaceae)

Fig (= Anjir, Tenatti, Simmayati) : *Ficus carica* L. (Fam. Moraceae)

Filberts (= Hazelnut) : *Corylus avelana* L. (Fam. Corylaceae)

Galgal (= Hill lemon, Sohlong) : *Citrus pseudolion* Tan. (Fam. Rutaceae)

Grape (= Drakhsha, Angur) *Vitis vinifera* L. (Fam Vitaceae)

Grapefruit (= Breakfast fruit) : *Citrus paradise* Macf. (Fam. Rutaceae)

Guava (= Anjir, Amrood, Peyara, Aim) : *Psidium guajava* L. (Fam. Myrtacrae)

Hog-plum (= Amra, Ambrella, Otaheite plum) : *Spondias cythera* Sonn. (Fam. Anacardiaceae)

Jackfruit (= Kanthal, Chakka, Panasa, Paka inchor) : *Artocarpus heterophyllus* Lam. (Fam. Anacardiaceae)

Jambhiri nimbu (= Rough lemon, Jambhuri, Citronelle, Jamir lebu) : *Citrus jambhiri* Lush. (Fam. Rutaceae)

Karna nimbu (= Karna khatta, Idnimbu, Karna orange) : *Citrus jambhiri* Lush. (Fam. Rutaceae)

Karonda (= Karamcha, Karekayee, Karamaroda) : *Carissa carandas* L. (Fam. Apocynaceae)

Lemon (= Pati lebu, Gora lebu, Baranimbu) : *Citrus medica* var. *limonum* Lush. (Fam. Rutaceae)

Litchi (= Lichee, Lichu, Lychee) : *Litchi chinensis* Sonn. (Fam. Sapindaceae)

Loquat (= Japanese plum, Loket, Nakota, Lakota) : *Eriobotrya japonica* Lind. (Fam. Rosaceae)

Madhuca (= Mahua, Mohual, Punam, Maul, Madgee) : *Madhuca indica* J.F. Gm. (Fam. Sapotaceae)

Mandarin organge (= Loose-skinned orange, Kinglove orange, Narangi) : *Citrus reticulate* Blanco (Fam. Rutaceae)

Mango (= Aam, Amra, Rasal, Amba, Mangga, Mangke, Amram, Mamide) : *Mangifera indica* L. (Fam. Anacardiaceae)

Mangosteen (= Mangoosta) : *Garcinia mangostana* L. (Fam. Guttiferae)

Monkey jackfruit (= Dephal, Barhal, Daua, Otomba) : *Artocarpus lacucha* Roxb. (Fam. Moraceaeae)

Musk melon (= Kharbuja, Kharmuj, Phooti) : *Cucumis melo* L. (Fam. Cucurbitaceae)

Olive (= Zaitoon) : *Olea europea* L. (Fam. Oleaceae)

Orange (= Same as orange)

Palmyra (= Total, Taar) : *Borassus flabellifer* L. (Fam. Arecaceae)

Papaya (= Pawpaw, Paita, Pepe, Pappalli, Parangimar) : *Carica papaya* L. (Fam. Caricaceae)

Peach (= Aaroo, Mandata, Takpo) : *Prunus perisca* Bats. (Fam. Rosaceae)

Pear (= Nashpati, Tang, French pear) : *Pyrus communis* L. (Fam. Rosaceae)

Pecan nut (= Pecan, Hicoria pecan) : *Carya illinoensis* Koch. (Fam. Juglandaceae)

Phalsa (= Dhamin, Phutifi, Parusha, Tadachi) : *Grewia asiatica* D.C. (Fam. Tiliaceae)

Pineapple (= Crown fruit, Ananas, Anaras, Kajudhakacha) : *Ananas comosus* Merr. (Fam. Bromeliaceae)

Pista nut (= Pesta, Pestalick) : *Pistacia vera* L. (Fam. Anacardiaceae)

Plum (= Alubokhara, European plum) : *Prunus domestica* L. (Fam. Rosaceae)

Pomegranate (= Darimba, Dalim, Anar) : *Punica granatum* L. (Fam. Punicaceae)

Pummelo (= Shaddock, Batabi, Chakorta, Pamparamazam, Papnasa) : *Citrus grandis* Osb. (Fam. Rutaceae)

Raspberry (= Rasbhari) : *Rubus ellipticus* Smith (Fam. Rosaceae)

Sapota (= Chiku, Safeda, Sapodilla) : *Achras sapota* L. (Fam. Rosaceae)

Sour lime (= Mexican lime, Key lime, Kagzi lebu, Nimbuka, Eloomechai) : *Citrus aurantifolia* Swingle (Fam. Rutaceae)

Strawberry (= Bhoomi berry, Estabhar) : *Fragaria* species and hybrids (Fam. Rosaceae)

Sweet orange (= Orange, Tight-skinned orange) : *Citrus snensis* Osbeck (Fam. Rutaceae)

Tahiti lime (= Persian lime) : *Citrus latifolia* Swingle (Fam. Rutaceae)

Tangor (= Tangor mandarin) : Hybrid of Mandarin orange and Sweet orange (Fam. Rutaceae)

Trifoliate orange (= Three-leaved orange) : *Poncirus trifoliate* Raf. (Fam. Rutaceae)

Walnut (= Akhrot, Krot, Persian nut) ; *Juglans regia* L. (Fam. Juglandaceae)

West Indian cherry (= Valouri) : *Malphighia punicifolia* L. (Fam. Malphighiaceae)

Watermelon (= Tarbuja, Tarmuj, Red fruit) ; *Citrullus lanatus* (Thumb.) Mats. and Nakai (Fam. Cucurbitaceae)

Appendix II

Conversion Factors

LENGTH

1 kilomere	=	10,000 metres = 0.621 mile = 1094 yards = 3181 feet.
1 metre	=	0.001 kilometre = 10 decimetes = 100 centimetres
1 mile	=	8 fuflong = 1760 yard = 5280 feet = 1.609 kilometres = 1609 metres.
1 yard	=	0.9144 metre = 9.144 decimetres = 91.44 centimetres = 914.4 millimetres = 3 feet = 36 inches = 2 hands (aprox.) = 4 bighats (approx.).
1 foot	=	0.3048 metre = 3.048 decimetres = 30.48 centimetres = 0.33 yard = 12 inches.
1 inch	=	0.0254 metre = 0.254 decimetre = 2.54 centimetes = 0.083 foot.

AREA

1 hectare	=	100 ares = 10.000 square metres = 2.4711 acres = 0.8237 *bighas* = 0.4114 *cottahs.*
1 square mile	=	640 acres = 259 hectares.
1 square metre	=	10.765 square feet = 1550 square inches.
1 acre	=	0.4047 hectares = 4840 square yards = 43,560 square feet = 3 *bighas* (approx.) = 60 *cottahs* (approx.)
1 square foot	=	0.0929 square metres = 144 square inches.
1 square inch	=	6.452 square centimetres.

Index